Spirit of the Environment

'This biodiverse (international) anthology of soulship could not have come at a more apposite time. As the technocrats in their self-styled first world decide that Mammon or at least the global economy is God. So there is no place for the other products of creative evolution let alone for those 'Heathen' who still attempt the life of subsistence farmers. Theirs is the dilemma of the Prince Bishops of the sacred grove at Nemi. They cannot afford to sleep for the greatest computer game of them all. The bulls and bears of hyperspace, goes on. Theirs is the Midas Touch.'

David Bellamy, *President of the Conservation Foundation*

'This is a welcome anthology, especially for its multicultural perspective, blending both indigenous and classical religious experience with more contemporary philosophical inquiry into the place of spirit in nature. The collection complements a North American emphasis on environmental stewardship and creation spirituality with independent and critical scholarship searching for the aesthetic, the sacred, and the spiritual in the human interaction with nature. The authors wonder whether a respect for life deepens into a reverence for life, and there is always the haunting question whether the ground under our feet links us with the Ground of our being.'

Holmes Rolston III, *Distinguished Professor, Colorado State University*

There are many people today for whom the most important questions about human beings' relationship to the natural world do not easily fit under such labels as 'scientific', 'ecological' or even 'moral'. For some people, the most central questions belong to religion – such as 'Was nature created by God for our use?, and if so, are there limits to our justifiable use of it?' For other, not necessarily religious, people, it is nevertheless the spiritual aspects of nature which should be a central dimension of environmental thinking.

Spirit of the Environment is designed as a textbook for courses in environmental studies, education, philosophy and religious studies, where study of the environment is not limited to examining the efficient utilization of the environment, but includes reflection on nature as the source and object of deep human aspirations.

David E. Cooper is Professor of Philosophy at the University of Durham.
Joy A. Palmer is Reader in Education and Director of the Centre for Research on Environmental Thinking and Awareness at the University of Durham. Both have published widely; they are the editors of *The Environment in Question* (1992) and *Just Environments* (1995), also published by Routledge.

Spirit of the Environment

Religion, value and environmental concern

Edited by
David E. Cooper and Joy A. Palmer

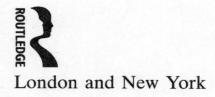

London and New York

First published 1998
by Routledge
11 New Fetter Lane, London EC4P 4EE

Simultaneously published in the USA and Canada
by Routledge
29 West 35th Street, New York, NY 10001

Phototypeset in Times by Intype London Ltd
Printed and bound in Great Britain by Clays Ltd, St Ives PLC

British Library Cataloguing in Publication Data
A catalogue record for this book is available from the British Library

Library of Congress Cataloguing in Publication Data
Spirit of the environment: religion, value and environmental concern/
    edited by David E. Cooper and Joy A. Palmer.
        Includes index.
        1. Nature – Religious aspects.   2. Human ecology – Religious aspects.
        I. Cooper, David Edward.   II. Palmer, Joy.
BL65.N35S69   1998
291.1'78362 – dc21   97–22308

ISBN 0–415–14201–6 (hbk)
ISBN 0–415–14202–4 (pbk)

Contents

Contributors

His All-Holiness Bartolomeus is the Ecumenical Patriarch of Constantinople.

Purushottama Bilimoria is Senior Lecturer in the School of Social Inquiry, Deakin University, Australia, and Visiting Lecturer in Philosophy at the University of Melbourne.

Stephen R. L. Clark is Professor of Philosophy at the University of Liverpool.

David E. Cooper is Professor of Philosophy at the University of Durham.

Greg Garrard is Lecturer in English at Bath Spa University College.

Rabbi Arthur Hertzberg is Vice-President Emeritus of the World Jewish Congress.

Fazlun Khalid is Director of the Islamic Foundation for Ecology and Environmental Studies.

Freya Mathews is Senior Lecturer in the School of Philosophy at La Trobe University, Australia.

Kay Milton is Reader in Social Anthropology at the Queen's University, Belfast.

Joy A. Palmer is Reader in Education at the University of Durham.

Martin Palmer is Director of the International Consultancy on Religion, Education and Culture (ICOREC) at Manchester Metropolitan University, and religious adviser to the Worldwide Fund for Nature.

Anne Primavesi is an author and was Research Fellow in Environmental Theology at the University of Bristol.

Kate Rawles is Lecturer in Philosophy at the University of Lancaster.

Richard Smith is Senior Lecturer in Education at the University of Durham.

Preface

This is the last in a trilogy of volumes on environmental thought which we have edited for Routledge. *The Environment in Question* brought together authors from geography, environmental science, philosophy and other disciplines, writing on topics as diverse as rain forests, obligations to future generations, and the ecological impact of tourism. *Just Environments* also ranged widely, though the focus was upon issues of justice and morality, such as our responsibilities towards animals and poorer nations, which our relationship to our environment raises.

There are many people, however, for whom the fundamental questions about our relationship to the natural world do not easily fit under such labels as 'scientific', 'ecological' or 'moral'. For some people, the central questions belong to religion. For example, was nature created by God for our use, and if so, are there any constraints on how we may use it? For other people, though they may be sceptical about the claims of any particular religious tradition, it is nevertheless – and in a suitably broad sense – the *spiritual* dimension of nature and our position within it which should be a main theme of environmental reflection.

It is not easy to state what this 'suitably broad sense' of spirituality is, and the contributors to this volume do not have a uniform understanding of it. But all of them would agree that a spiritual outlook on nature is to be contrasted with an instrumental or pragmatic one. All would concur in Charles Taylor's distinction between two competing outlooks in modern 'ecological politics': an attitude of 'instrumental reason', for which the solutions to environmental problems are 'technical', *versus* a recognition that 'we are part of a larger order' from which our own life 'springs and is sustained', with the consequent imperative that 'we be open to or in tune with nature' (Taylor 1989: 384). In the 'suitably broad sense', those who experience a significance or purpose in nature which we should honour, or who find in it a source of aesthetic inspiration, are viewing nature 'spiritually'. Hence, in this volume, there are chapters on art and nature, nature as a source and object of wonder, the natural world as an 'ecological self', the Romantic perception of a nature imbued with meaning, and the implications of the Gaia hypothesis, as well as chapters on the positions of the major religions, east and west, and on other important religious traditions, notably pantheism.

As in our previously edited volumes, no attempt has been made to lay down a uniform approach by our contributors. They come from a variety of disciplines – philosophy, theology, comparative religion, education, literature, social anthropology. Some write as commentators on, or critics of, traditions; others are proposing new ways in which we might view our relationship to the natural world. Some reflect on central concepts which pervade our thinking about that relationship, such as the very notion of knowledge or understanding of nature; others draw on their empirical research to challenge received opinions about, say, the spiritual bonds between traditional peoples and their environments, or the role of spiritual considerations in children's attitudes to the world about them.

Taken together, the chapters form a richly variegated book suitable for use as a text on courses – be they taught in departments of environmental studies, education, philosophy or religious studies – where 'ecological politics' is not confined to examining the technical means whereby we may more efficiently utilize our environment, but extends to reflection on nature as the source and object of some of our deepest, and most distinctively human, aspirations.

Joy A. Palmer
David E. Cooper

1 Indian religious traditions

Purushottama Bilimoria

INTRODUCTION

The Indian religious traditions are intertwined with equally disparate cultural, social, linguistic, philosophical and ethical systems that have developed over a vast history, compounded with movement of peoples, foreign interventions and internal transformations in structures and identities experienced over time. How does one then begin to talk about environmental values and concerns in the Indian religious traditions? Well one can, albeit randomly and selectively; and so this essay will be confined to tracing the contours of certain highlights and tensions in the traditional approaches to the question of the environment. Of special significance will be the Brahmanical-Hindu, Jaina and Buddhist traditions, in their ancient to classical modalities, concluding with some contemporary responses to the supposed impact, or lack thereof, of traditional perspectives on ecological problems facing a rapidly modernizing South Asian nation state, from Gandhi to Bhopal and after.

Even before the Brahmanical order took firm root in greater India, there are records from incomplete archaeological findings that suggest a major civilization of the Indus Valley (in a sprawling region encompassed by the Punjab, Sind and present-day Pakistan and Baluchistan), which peaked around 3000 BCE, where a close symbiosis between nature and the Dravidic people appears to have been prevalent (Wheeler 1979: 1, 84). The major cities of the Indus civilization, namely Harappa and Mohenjo-daro, with their imposing civic edifices, mudbrick and timber dwellings complete with baths, extensive drainage and sewer systems, give the impression of being exceedingly carefully designed. The architecture as well as farming practices give evidence of structural harmony with surrounding and climatic conditions that would optimally conserve natural resources, prevent deforestation and appease the gods who were little more than personified symbols of human dependence upon the energies of nature.

Some elements of the religious and cultural practices from the Indus period and other indigenous (especially aboriginal) communities continued into the subsequent Vedism phase, which began with an influx of Aryans or 'Noble People', a tribe of pastoral nomads from somewhere in Central Asia who settled on the plains of the Ganges in the northern part of the subcontinent around second millennium BCE. Their agrarian culture, so much dependent on the forces of nature, is reflected in the repertory of hymns, the earliest of which are known as *Rig Veda*. The oral tradition and the Veda would have to be among the earliest record of ruminations on nature in India. The Vedas (from Sanskrit *veda*, 'what is known') gradually grew into a huge canonical body of recited and memorized 'texts' that both eulogized and appeased the forces of nature and higher planes of beinghood depicted as gods. These were used in liturgical sacrifices and elaborate rituals supported by the chanting of sacred mantras or coded formulaic syllables, although their distinctive philosophical import remains largely hidden.

EARLY INDIAN ETHICS AND OUTLOOK ON NATURE

It is perhaps a remarkable feature of the Indian tradition that from its very early beginnings ethical ponderings were never too far from the overwhelming awareness of nature, in as much as 'forms of life' were derivative of or entailed by a particular outlook on nature of which the human being, as other species or sectors of beinghood, was seen as a constitutive, at times lost, alienated or anomalous, perhaps even an out-rageous or offending, part. In their moral judgements, the early Indian people placed on the side of the 'good' such values as happiness, survival, courage, health, joy, calmness, friendship, knowledge and truth; and on the side of 'bad' more or less their opposite or disvalues – misery, suffering, sickness and injury, death, infertility, pain, anger, enmity, ignorance or error, untruth (Bilimoria 1991: 44). These normative values were not restricted for human well-being alone, rather they were universalized for all sentient beings and inanimate sectors: spirit-spheres, gods and the faithfully departed; the biosphere, animals and plants; and the broader biotic universe, inanimate realms comprising the elements, stones, rocks, earth-soil, mountains, waters, sky, the sun, planets, stars and galaxies to the edges of the universe (this and other possible ones).

The principle guiding this outlook was that the highest good is to be identified with the total harmony of the cosmic or natural order, charac-terized in the earliest religious texts as *rita*, which we could render for now as the natural law: this is the creative purpose or telos that circum-scribes all sentient behaviour and every movement, from the stillness of the deep-sea water to the invisible vibration of the sub-atomic particle. The social and moral order is thus conceived as the correlate of the natural order. The vast universe was not strewn about in random chaos, but had an inner order, a unity with an inexorable law and purpose (*rita*) that governs the working of both the macrocosm and microcosm

(Dandekar 1979: 15). This is the ordered course of things, the truth of being or reality (*sat*) and hence the 'Law'. *Rita* determines the place, entitlement, function and end of everything. But *rita* is too subtle for the undiscerning eye, and its originary promulgation occurs mythically with the dismemberment of the Cosmic Person (*Purusa*) performed by the gods. Following the differentiation of the cosmos, numerous gods, often in a spirit of competition, would claim the title of the supreme enforcer of the Law. This indeed coincides with shifts in the substantial environmental conditions of the Aryans on their further migration towards the seven big rivers (Saptasindhu). Thus, Indra, a human hero who evolves to become the chief of gods, is extolled for his command over the arid forces of nature, especially the thunderstorm and thereby the refreshing of the earth with rain. The other gods who variously regulate different aspects of the biotic community are perceived as working in unison with the mind of Indra. He claims to have released the Sun from its concealing darkness, and set the solar-disc on its proper course in the sky, making it shine bright so as to give energy to all gradations of sentience and nature – animals, trees, waters, rocks, moon, and so on; and in turn the Sun-god, Savitri, looks over to see that all other gods live according to *rita* – the harmonious interplay of all the elements as forces of nature.

Later on, in *Atharva Veda* (VII.24.1), Truth (*satya*) is identified with *Dharma*, as the Law that governs all beings. In the moral sense, Truth stands for integrity, living by truth and not falsehood. *Rita* as an aspect of natural law is here given a deeper ethical nuance. This requires active and positive abidingness with things as they are even as each thing strives towards the realization of its own intrinsic good in the larger scheme of things. For everything in the universe in a deep sense is thought to have its own worth and is therefore morally significant. It is in accord with this moral understanding that each thing, or species, consciously or unconsciously, individually or collectively, would strive to realize the deeper truth in their own uninimical way. Thus, when in accord with the Law the rains break, the 'fountains' that 'bubbling, stream forward' are said to be 'young virgins skilled in Law' (*Rig Veda* IV.19.7), and the microamoeba (primary cells) fight off opposing cells in order to bring forth life. Thus a distinction is assumed between good interests and untoward interests, and the moral standing of the respective species or 'island' is determined in accordance with what particular interest or set of interests the species best serves.

In ecological terms, the Vedic hymns interweave a number of insights, from a primitive conception of a unique all-being (or non-being) of which everything is a part, to the more complex idea of everything being a part of a unity which is also in everything or in every part that is constitutive of the unique whole. In other words, the Vedas speak of the uncanny unity of creation and, more significantly, the mysterious interconnectedness or codependence of everything on everything else. Each thing, element and species or bio-organism – which can be characterized as having the mark of beinghood – has an interest and purpose to fulfil in the larger scheme of things. It is this that makes each thing 'sacred' and therefore worthy of moral consideration, by human beings and the gods alike.

The ancient people recognized that they could neither control the whole of nature nor interfere unduly in its order and processes to seize control of all its varied functions; that, if anything, they needed the cooperation of the benign and harsh elements alike, be these the ravaging sails of the wind, the bursting of the waters, the quake of the earth, the fire of the forests, the wild beasts and pests on the fringes of dwellings, the darkness of the night, the stubborn seasons, and so on. Only after understanding the system and much sacrifice, that is, appeasing of the forces of nature and the spirits in command beyond, could they hope to benefit from the bounty and goods provided by nature, or design wheels and other instruments for extracting natural products, dictated by needs rather than want and greed. Rituals helped prepare plants, herbs and other healing products to restore health and rectify breakdown of the Law. Strict equilibrium had to be maintained in the internal environment as it was the Law in respect of the external environment too. The ecological framework in a broad stroke was formulated in terms of the proportionate combination of matter (substance, atomic entities) and energy (variously imaged as the spirit, breath, speech, vibration, anima, pneuma). Competition over the resources of nature can deplete the energy levels and create an imbalance in the polar relations. The human being has no prelapsarian claim of dominionship over nature. A classical (Benthamite) model of utilitarianism which measures pleasure (or gain, benefit, the good) in terms of human interests alone could not have been thought of in this context even as a theoretical or formal possibility. The interest of the 'deep whole' or species in the broadest possible sense cannot be overlooked or unreasonably compromised.

However, some competition within nature represented in terms of struggle and tension between and among individuated forces signifying matter and energy is not ruled out; indeed, this could be a healthy crisis point and provide incentive for growth and flourishing of the natural world and towards overcoming malignant matter, 'evil' spirits or bad omens that hinder progression. But competition with nature can lead to disastrous consequences as well. The later Vedas, especially the Ayurveda section, demonstrate profound knowledge of biodiversity, the interrelationship between living species and the environment, the need to maintain natural dynamism, the right ways of handling plants and trees, native flora and fauna, or the price one pays for transgressing the ecological principles. The attitude was invariably one of mutual respect, reciprocity and caring for other (non-human) subjects of the land. Appropriate belief-states along with commensurate rituals were developed that reinforced and continued this symbiotic relationship. The symbolic ritual act of appeasing the 'soul' of the tree before removing it to clear space or land for human habitat or use is indicative of the respect afforded to the natural world. Recycling was a highly valued practice in traditional India, recognizing that certain trees and plants do not even as much as tolerate wastage of their fallen branches, twigs, seeds and flowers (they may regenerate into another plant or be self-composted).

However, there are reservations about the traditional account. First a general point: it should be pointed out that despite the rhetorical strokes that sweep across the universe or cosmos, much of the ecological concerns and activities were confined to the more or less perceptible reaches of the surrounding or local ambience. At the farthest edges of the dwelling villages and agricultural terrains lay dense tracts of forest and jungles which were almost impenetrable (except by indigenous tribes, thugs or an attacking army), beyond which one had little recourse to be concerned about how the 'alien' groups organized their lives, tilled their land, or disposed of the departed, and so on. The chief imperative was to get one's own house, as it were, into some semblance of order and harmony; the universal appeal or applicability may follow later, gods willing. Second, one too often overlooks the negative effects of sacrifice, as this entails killing of animals, usually from the best of a breed, and sometimes this becomes a widespread practice as superstition sets into a culture. Third, it neglects the expropriations and amassing of power via Brahmanical or other upper-caste privileges which in the past have led to the deprivation of the basic necessities of life and share in the goods of nature on the part of lesser groups, classes and sectors of the population, women included. Fourth, in the master–slave ideology that ensues, the exploitation of human labour extends to the exploitation of animal labour and competition for natural produce, which, as history has attested, results in wholesale colonization of vast tracts of natural landscapes. Fifth, in their eagerness to cultivate, to increase production and accumulate goods, rulers and landlords fall short of careful planning and do not take adequate steps against despoilment and damages to natural surroundings; in other words, they have no environmental programme as such. Sixth, the zeal for expansionism instigates rivalry and even warfare between neighbouring kingdoms, provinces and states, causing much harm to the buffer forest zones and to each other's settlements.

BROADLY CLASSICAL

Following the pattern of pre-classical religiosity, Hinduism developed a strong moral ethos (*dharmaśamsanta*) which to a large extent superseded the earlier (Vedic) view from the heavens (or of the gods) by a view as if, 'from nowhere', that is, from no one particular subjective position (whether divine or human). Here the moral concept of *dharma* emerges as a much more abstract, authoritative and autonomous notion, but with the same normative strength that the ontological and cosmological conceptions had earlier served. The universe is seen as a most meaningful and principled moral order: human beings have a responsibility, indeed a duty, to help sustain this world thus rendered morally significant or 'deep'. (The Sanskrit root '*dhr*' means to sustain, uphold, support.) The difference from similar sentiments built around the idea of *rita* as an 'eternal order' or alternatively as fixed principles is that here the moral content is deepened, in that it is much more concrete and better defined, it is normative, at times legalistic or systematic, issuing in elaborate proscriptions, precepts and rules, ordinances and statutes, which are written down in a great many texts, including personal ethical or moral manuals as well as social and political treatises, such as The *Dharmaśastras*, *Arthaśastras*, and the *Mahābhārata*. *Dharma* comes to designate a variety of moral terms – norm, virtue, righteous duty, responsibility, entitlement, justice, truth (in conduct) – and there continues to be much debate and hermetic anguish over its exact nuance and application. But its universal appeal is perhaps in its calling to preserve the organic unity of beinghood, to render justice where justice is due, and to minimize the burden of *karma*, which reflects a universalization of the basic tenet, 'as one sows so one reaps' (Bilimoria 1995). The rule of *karma* does not discriminate between humans and non-human life-forms (amoebic, individual or whole ecosystems) for everything has value and is an end in itself. In ethical terms it demands a deontological disposition in one's conduct, but its own internal calculative system heeds to consequences, good and bad, and to any excesses of utilitarian or even prudential exploitation committed on the part of one species over another or the others (see Midgley 1995a: 97). And duty, it follows, is here cast neither within the theoretical frame of contractual obligation nor as a necessary response to corresponding rights; duty is performed, as it were, for *dharma*'s sake: the sense of responsibility is *sui generis* a (moral) relative absolute. Linking ethics to the parameters of certain religious cosmologies would entail that there are some duties which are mandatory.

One of the cardinal duties and therefore values to be developed in the *dharma* tradition (in the shadow of the proto-yogic descendants, Jainism and Buddhism) was that of general non-injury. The most refined expression of this value is represented in the great epic of the *Mahābhārata* (*circa* 100 BCE to 200 CE). Much moral development proceeds through organizing and placing constraints on the otherwise presupposed liberties of human life. Of chief concern is the impact that one's action, pursuits and conduct might have on the other. In this regard, non-injury or non-violence is prescribed unequivocally.

The *Bhagavadgītā*, which is a book within the great epic, provides a quasi-philosophical grounding for the values extolled in the *Mahābhārata* and is more decisive in its ethical pronouncements. It is for this reason that the *Gītā* (for short) has had a profound impact on modern Hindu-Indian thought and is drawn upon obliquely in Western ethical and eco-logical deliberations as well (Gandhi 1962; Naess 1990: 194; Jacobsen 1996: 231–3). Two most commented upon verses in this context are the following:

> The one whose self is disciplined by yoga
> Sees the self abiding in every being
> And sees every being in the self;
> He sees the same in all beings.

> When one sees pleasures and pain of others
> To be equal to one's own, O Arjuna,
> He is considered the highest yogin.

> (*Bhagavadgītā* VI.29, 32)

Several commentators, including Śaṅkara, have observed that the feeling of pain is universalized so as to derive a principle of empathy and non-injury. Śaṅkara characteristically commented that one who sees that what is painful and pleasant to himself is painful and pleasant to all creatures will cause no living beings pain, and that he who is non-injurious is the foremost of yogins (Śaṅkara 1976: 198–9; Bilimoria and Hutchings 1988: 366). Self-realization in the *Gītā* takes due cognizance of the moral principle of *lokasamgraha*, the well-being of all peoples. The world of living things is brought together in a process governed by moral cause-effect relationships and it makes it imperative for each being within it to respect the autonomy, the interests and destiny of the other, and ultimately to find a way out of the cyclic implications of this process.

ŚRAMANA TRADITIONS

This broad moral characterization is true not only of Hinduism, but the parallel – or by some accounts alternate – traditions of Jainism and Buddhism that also developed extensive moral theories and cultures, which we shall now explore for their ecological ramifications.

Buddhist and Jaina religious philosophies (along with yoga-asceticism from the time of the Indus civilization) are said to belong to the Śramanic lineage or the proto-yoga renouncer tradition The *Śramana*, very gener-ally, had its origins in lonesome ascetic pursuits by socially withdrawing (or 'drop-out') yogis, that is, adepts of certain esoteric practices who became disenchanted with the prevailing ritualistic or materialist human environment, giving way gradually to a monastic and more formally speculative system of ordering life. The systems evolving out of this traditional lineage were looked upon as being somewhat eccentric, non-mainstream and heterodox by the powerful Hindu-Brahmanical ortho-doxy. Nevertheless, Jainism and Buddhism both grew to gain considerable strength and following within India; and while the popularity of Buddhism shifted further North to Tibet and eastward to Southeast Asia and the Far East (China, Japan, Korea), its influence on Hinduism was quite extensive. (Indeed, in large measure the challenge of Buddhism was responsible for the gradual erosion of the Brahmanical orthodox stronghold giving way to a more broad-based, though still caste-ordered, popular Hinduism in the post-classical phase.)

Jainism owes its name to the term *jina*, meaning one who conquers attachment and overcomes pain. The prominent Jina who helped give a more formal shape to the order and systematized the teachings of an older group of Jinas was Mahāvīra (*circa* 500 BCE), possibly a contemporary of the Buddha. The basic philosophic belief of the Jainas is that every entity in the world possesses *jiva* or a sentient principle, and there is a countless number of *jivas*, whose distinguishing feature is consciousness along with vital energy and a pleasurable disposition. The suggestion is that consciousness is continuous and nothing in the universe is without some degree of sentience at varying levels of conscious and apparently unconscious existence, from its more developed form in adult human beings to invisible embryonic modes at 'lower' animal and plant levels. (Here sentience is not determined merely by pain-pleasure responses, as some psychical activity may continue to occur etherically or subconsciously or at unconscious levels as also under naturally disposed comatose and anaesthetized conditions.) The sentient principle *jiva* subsists in a contingent relation to the quantity of *karma* it has accumulated through its activity, volitional and non-volitional. If *karma* can be prevented and exhausted, the bondage of cyclic process of existence can be arrested, and the sentient being would achieve full self-realization. Since this requires much discipline (of self-control and renunciation) and the process is long and arduous (extending over several embodiments and re-deaths), each sentient being has to act in accordance within its relative level of bondage and limited freedom. The Jinas remain the sagely exemplars, while 'lesser' sentient beings, at least in terms of practical ethics, are considered immune from moral frailty, though they are not devoid of moral value in their own right by this theory. The cardinal disciplinary codes highlighted in Jaina practical ethics are: *ahimsā* or non-injury, *satya* or truthfulness, *asteya* or not stealing, *brahmāchārya* or sexual restraint, *apigraha* or non-possession.

Recent scholars have gallantly extolled the virtues of non-injury or non-violence, *ahimsā*, in part because the Jaina ethic of non-injury is as much part of a regime of internal discipline as it is of external conduct or behaviour towards others. They also tend to endorse the operative cosmology of the Jainas as 'perhaps [being] the most sympathetic to an ecological world view' even while recognizing that the basic teleology of the Śramanic traditions is aimed towards transcendence of the self from the constricting human conditions of desire and attachment (Chapple,

1994: 9–18). The second point to note is that the Jaina ethic of non-injury and a compassionate regard for others (insects, microamoebic entities, animals, human beings, gods and spirits) finds its support on prudential grounds, for doing harm to other beings will result in more negative *karma* for oneself! Thus the ultimate justification for all ethical practices is that they should raise the moral stature of the practitioner; if derivatively, perhaps unavoidably given the interconnectedness of all *jivas*, it raises the moral profile of the community (in the broadest biotic sense), then this is all the more reason for persisting with it. Some writers, however, would argue that such virtues as *ahimsā* have intrinsic value and that their justification lies in their being derived not from objective facts (such as 'all life has sanctity'), but from some experience which is self-evident. What is 'right' is in harmony with this experience. *Ahimsā*, in their view, is an experience related to the occurrence of pain and suffering among living beings and is universalized for others from one's own experience of pain. *Ahimsā* stands as the 'good' to which other values tend (Songani 1984: 243; Bilimoria 1991: 53). Hence it follows that if there is to be a clearer articulation of Jaina environmental ethics it too would strive to be autonomous and normative, admitting the possibility of objective value, of which *ahimsā* would seem to be the most significant and distinctive feature.

The picture is more or less consistent with the case of Buddhism as well, although Buddhist ethics proceed on a broad-based naturalistic stance, as Buddhists would concur that certain types of fact are relevant as support for moral considerations (de Silva 1990: 18; 1991: 63). One such general fact with which the Buddha began his teachings is that there is suffering, for the human condition and the surrounding state of affairs confirms this truth, not least the contingency of existence (birth and death) and the impermanence of all things, good and bad, big and small, here as elsewhere. The appropriate moral response is to minimize suffering and pain as best as one can and to overcome suffering or unsatisfactoriness, both by understanding the causes of such existential and other ailments, and by alleviating the suffering of all forms of life. Its ethics cover human behaviour in relation to all living beings and it underpins certain basic virtues, particularly of the benevolent kinds, more specifically, compassion, love, kindness, sympathy, empathy, equanimity and joy in the other's happiness. It is said that human beings are capable of infinite amounts of compassion, generosity and gratitude, and that all creatures, great and small, should be the subject of our moral sensibility (Dalai Lama 1996). The Buddhist codes of ethics are similar to the Jaina ethics, with much emphasis placed on self-control, abstinence, patience, contentment, purity, truthfulness and right attitudes. The treatment of animals and plants in accordance with these principles finds ample references in Buddhist texts, from the earliest monastic codes to the development of Ch'an or Zen Buddhism in China, Korea and Japan.

Nature as a whole is not looked upon as antithetical to human needs; rather, everything in nature is capable of making a contribution towards overcoming suffering and the final spiritual end towards which human beings strive. The Buddha's teachings include tales of acts of generosity on the part of animals towards human beings, and the reciprocal compassion which humans are advised to direct towards other life forms. Buddhist societies evolved with this moral self-consciousness, and the great emperor Aśoka, after his conversion to Buddhism, institutionalized care and welfare towards animals, as the following edict tells us:

> Here no animal is to be killed for sacrifice. . . .
>
> Formerly in the Beloved of the Gods' kitchen several hundred thousand animals were killed daily for food; but now at the time of writing only three are killed – two peacocks and a deer, though the deer not regularly. Even these three animals will not be killed in future. . . . the Beloved of the Gods has provided medicines for man and beast. . . . medicinal plants . . . [R]oots and fruits have also been sent where they did not grow and have been planted along the roads for use of man and beast.
>
> (*Sources of Indian Tradition* 1988: 144–5)

The verses demonstrate that rights and protection of certain liberties of animals have been recognized in Buddhism. Many Buddhist monasteries across East Asia as well banned the cooking of animal flesh as this involves the killing of animals, with or without direct intentionality of consumption. Buddhist environmentalists are active in modern-day Sri Lanka in their efforts to preserve the lush beauty of the island state from despoilment through extensive technological development and the ravages of an ethnic war that has escalated there in recent decades. They, too, can be said to be continuing a practical environmental ethic fostered centuries ago after Buddhism was brought to Sri Lanka.

Likewise the arrival of Buddhism in Tibet in the seventh century engendered a nationwide programme for the preservation of the heavenly–natural oasis that remained a mysterious land for much of the outside world. The ruling lamas proscribed injuring and killing of animals, big and small. The moral practice of showing respect for all nature became a way of life for the Tibetans. Even though Tibetan Buddhist metaphysics continued the influential Indian Buddhist doctrine of the absence of self-nature or intrinsic existence of properties and substances alike, proclaiming thus the emptiness of all things, its moral framework paradoxically gained strength from this standpoint, on three counts:

1 Moral properties such as those of the good, compassion, and loving kindness or respect, by no means absolute, have solid presence (contingently supervenient on 'emptiness', of course), in as much as

human interaction and communication or ethical life generally presuppose these properties.

2 A pluralistic ontology that has fair regard for members within it without privileging any particular species easily gets translated into a non-anthropocentric respect for biodiversity.

3 The religious–soteriological 'end' requires certain self-motivated ethical practices and norms, including restraint on desires, meditation on the limits of the ego-self, altruism based on the moral properties of reverence and deep (but not condescending) compassion for all living and non-sentient beings. In other words, the normative constraints on monks, nuns, lay people, farmers and nomads, too, underscored concern for the environment.

The Buddhist ethic of living in harmony with the earth accordingly pervaded all aspects of the Tibetan culture. Perched on the 'roof-top' of the world or on an altitude shared with the Himalayas, Tibet's environment was recognized as being crucial to the stability of ecological environs and crop cycles in much of neighbouring Asia. For instance, the ten or so major rivers that wind through Asia feed off the river valleys and smooth glacial icescapes of Tibet; the monsoons sweeping through South and Southeast Asia rely on Tibet's abundant natural vegetation and dense forests. Its wildlife and natural animal sanctuaries maintained a natural equilibrium and contributed in different ways to the enrichment of the environment, providing manure for controlled husbandry and organic re-vegetation, as well as fuel (from yak dung), and so on (Oxley 1996: 1, 2).

However, after the Chinese occupation of Tibet around 1950, the situation has dramatically altered: massive deforestation, land erosion, pollution of rivers, depletion of resources, excessive killing of animals, and general degradation of the environment appear to have become the norm. The information is sketchy, video recordings or testimonies smuggled out of Tibet are not always reliable. But official Chinese obfuscation adds to the suspicion. Observers lament that the spirit of Buddhism is being crushed in Tibet and claim that the environmental damage will continue until such time as the patrons of Buddhism, namely Tibetans with their refugee spiritual–temporal head, the Dalai Lama, are returned full cultural control and self-determination of the country. This shows the faith that some people have in at least one field of traditional wisdom, in regard to the environment.

CONCLUDING REMARKS

Traditional Hindu, Jaina and Buddhist environmental values and concerns have continued to influence the discourse and aligned practices of environmentalism in much of South Asia. One of the most successful and well-noted applications of the Indian ethic of non-injury emerged with the non-violent struggle led by Mahatma Gandhi in the first half of this century. Gandhi was adamant about the need for such an ethic in our treatment of animals as in our behaviour towards each other and towards other human beings (Gandhi 1959: 34–5). He followed a strict regime of vegetarianism (bordering on vegan practice, except that he accepted goat's milk). Unlike Jainas, however, whose practice of *ahimsā* could best be described as a form of passive self-restraint, Gandhi turned *ahimsā* into a dynamic force, informed by truth (*satya*), that proactively engages in the promotion of non-violence and achieves its various social-political goals through activities grounded in non-violence, which becomes the outward symbol of the inner 'truthforce' (*satyagraha*). A spectacular environmental movement called the Chipko (from an Indian vernacular term meaning 'cling on to', which describes an unrelenting embracing of the trees to prevent environmental destruction through human intervention) was directly influenced by Gandhian environmental awareness programmes and led by Gandhian *sarvodaya* (welfare-for-all) workers on the principle of non-violent resistance (Weber 1988: 24). Nevertheless, Gandhians by no means believed in complete biospheric egalitarianism and permitted small-scale or modest introduction of 'soft' technology supplemented heavily with hand-crafting and cottage industries localized to village economy.

Another case which drew worldwide attention where similar non-violent resistance tactics have been used to raise awareness of environmental concerns is the Narmadī Dam project in south Gujarat. Environmentalists have constantly argued that damming the river would cause immense damage to surrounding land which would also lead to the dislocation of masses of tribal people who have lived in the vicinity with good regard for their environment for countless generations. The intensive protests provoked the World Bank to withdraw its share of promised funding. There are numerous other grassroots groups and movements that invoke traditional wisdom and practical ethics in their expression of resistance to and concerns for radical transformations of the local environment. There is great apprehension that these interventions serve the technocratic interests of upper classes, the middle-managerial classes or the national, or, as is increasingly the case, multinational corporations and mega-media tycoons who have no understanding of or sympathies for local conditions, customs, habits, attitudes and the underpinning cosmologies or philosophies. Rural development and alternative technology programmes have been helping villagers and farmers to construct, for instance, free-standing smoke-less ovens and mudbrick dwellings, and to utilize non-toxic organic fertilizers in well-irrigated farmlets for their produce. Schools and colleges are established with the help of non-government groups to explore and promote safe ecological practices. Tribal groups have been encouraged to preserve the wild bushland, to curtail excessive use of wood for fire-cooking, and to develop new kinds of

technology for dealing with local conditions while resisting the technologies and wares brought in by eager profit-driven urban and corporate
enterprises.

However, despite the great wealth of wisdom and inspiration afforded
by traditional teachings and cosmological blueprint that underscores
strong ecological values, a number of writers and critics on India have
mild to strong reservations about the relevance of such traditional
approaches. This cleavage surfaced in the aftermath of the Bhopal inci-
dent in 1984. The Union Carbide chemical plant, which had been ill-
maintained for some years, unleashed thousands of tons of poisonous
fumes and chemicals in the atmosphere which killed and irreversibly
handicapped thousands of people. As with Chernobyl, the enormity of
the Bhopal catastrophe could not even have been imagined by traditional
wise-men, and so one questions whether tradition, including perhaps
Gandhian minimalist industrial programmes, could have ever alerted and
therefore prepared society for such an environmental holocaust. The
naturalistic fallacy notwithstanding, if the facts were not there facing them
in their eyes, what motivations or triggers would the ancients have had
for pondering on correlate values that would be necessary to contain
or deal with the facts? The world has changed and the challenges of
industrialization, modernity, globalization and a rapidly expanding liberal
economy present us with very different sets of circumstances and contexts
that require quite different sorts of responses on the environmental front.
Are there any resources left within the traditional framework to combat
the modern consumer model which has all but disrupted the traditional
agricultural practices and all kinds of unities? asks one of the best-known
Indian activists and environmentalists (Shiva 1988). But Shiva for one
does not underestimate the contribution traditional or pre-modern sensibilities can make towards fostering a 'post-modern' response in the terms
of an integrated, holistic view of both humans and their environment
(Shiva 1988). She more recently supported a nationwide campaign against
'plant variety' rights claimed by Western multinationals under intellectual
property and international patenting accords, to which such countries as
India, several South American states and Australia have been persuaded
to become signatories. This latter move is seen by environmentalists as
acting against biodiversity and the right of each people to control and
maintain their local ecosystems within the means and wisdom afforded
by traditional or customary practices and modern-day urban pressures.

Still, there are critics, such as Ram Chandra Guha and Chapple, who
suggest that a too one-sided focus on traditional patterns of ecological
thinking and attitude detracts from the need of the hour, which is an
active and practical initiative for addressing local and specific or particular
instances of environmental abuse, of degradation, and violation of agreed-
upon international memoranda for the protection of living and non-living
species. Thus, Chapple has argued that although 'the integrated reality of

village economy, as espoused in the case of the Brahmanical traditions, certainly sustained agrarian India for millennia, and although tribal peoples today continue to eke out a sustenance existence, neither model bears direct relevance for the burgeoning urban life that hundreds of millions of people in India have embraced in the past few decades' (Chapple forthcoming). He concludes on a sad note: 'Unfortunately, both models suffer a platitudinous hollowness and, I am afraid, will fail to capture the imagination of precisely the sorts of people who stand to commit the greatest infractions against the ecological order, the people throughout South Asia who feverishly are buying cars, building condominiums, and filling their flats with prepared foods and plastics.' Perhaps Chapple is echoing the oft-made charge that environmental ethics lacks a sustained political ideology and programme (Sylvan and Bennet 1994). But his own alternative to the 'shallow' ecology from the hoary past verges on rekindling Grandhian suspicions of the virtues of technological-consumerist largesse and on deepening the Jaina ethic of non-violence to 'animals, earth and self' in a reinterpreted practical ecological ethics so as to accommodate current (and future) environmental concerns.

The suggestion is that there are indeed resources within the traditional systems – Yoga, Jaina, Buddhist, Hindu, Sikh, Islamic, Christian, Gandhian, all of which have helped give shape to a modern, secular India – to increase awareness of environmental concerns and to instigate the extension of ecological values and model practices to the plethora of environmental problems facing Indians, as they do most human beings in other parts of the world. This is a laudable suggestion and one with which a number of environmental thinkers are likely to agree, or, if they disagree, at least will debate.

2 Chinese religion and ecology

Martin Palmer

To enter into an understanding of Chinese religions and their attitude towards nature, it is necessary to begin to comprehend the Tao. At one level it is simply a word which means the path, the road, the ordinary name for a street. Yet, at another level, it is the term used to describe – by Taoists and Confucians alike – the ultimate Way of the Universe. For example, the classic text, the *Tao Te Ching* (*The Way and its Virtue*), written somewhere between the fifth and fourth centuries BCE and revered as a sacred text by later Taoists, has this to say:

> The Great Tao goes everywhere
> past your left hand and your right –
> filling the whole of space.
>
> It is breath to every thing, and yet it asks nothing back;
> it feeds and creates everything, but it will never tell you so.
>
> It nurtures all things
> without lording it over anything.
>
> It names itself in the lowest of the low.
>
> It holds what it makes,
> yet never fights to do so:
>
> that is why we call it Great.
>
> Why? Because it never tries to be so.
>
> (Kwok, Palmer, Ramsay 1994: 103)

The Tao is that which brings life to all life – indeed, the Tao is life in all its forms. The true nature, the innate nature of all life is Tao and the Tao is that which is natural. The Tao is not a deity. The Tao simply yet gloriously is how life is and is the innateness of being.

The fourth-century BCE philosopher and Confucian scholar, Meng Ke, known in the West as Mencius, expresses this well in a famous text which shows that deforestation, with its resulting environmental problems, is no new thing!

The trees on the Ox Mountain were once beautiful. However, because the mountain is on the borders of a great state, they were cut down with axes and saws, so how could they retain their beauty? Yet they continued through the cycle of life and the feeding of the rain and the dew to put forth buds and new leaves. But the cattle and goats came and browsed among the trees and destroyed them. This is why the mountain is now bare and stripped. People look at it and think this is how it has always been. But this is not the true nature of the mountain.

(Author's translation of the *Book of Mencius*: Book VI, part A, section 6)

The concept of the true nature, the innate nature of all things, lies at the heart of the Chinese philosophical understanding of reality. All things have their innate nature and violation of this innate nature means violation of the Tao – the underlying innateness of all being.

The exploration of the concept of Tao lies at the heart of two of the three great religious traditions of China.

Taoism takes its name from the Tao. Its roots lie far back in China's religious pre-history. The earliest religion of China was shamanism, a world view in which this physical world is but one of two worlds. The other world, the more significant world, is the spiritual world. The shaman is someone who in a trance can pass between the two worlds, taking and bringing messages and being imbued with the power of the spiritual world for healing and foreknowledge of the future. The earliest myths of China are all shamanic with the great rulers being shamans who can change their shapes and become the totem animals through which they can more easily move into and out of the spirit world.

As China moved towards settled kingdoms, the shaman rulers gave way to warrior kings and hereditary lineages and the shamans became the priests of the kingdoms. Their power, therefore, was already waning by the time the Confucian political and philosophical world view began to develop around the sixth to fourth century BCE. The Confucians, as will be seen more clearly below, look for a structure and an order to life and if they don't find one, then they impose one. This runs contrary to the eclectic and spontaneous nature of spirit mediums. By the beginning of the second century BCE, the shamans had been pushed out of the court and marginalized. But their influence with ordinary people never waned. However, it underwent a transformation. In the first and second centuries CE, many shamanic ideas and practices re-emerged from among the ordinary people in the form of a new religious movement – Taoism. Taking its philosophical concepts from a mixture of shamanic practices, and ideas contained in the classic writers about the Tao such as Lao Tzu (the *Book of the Tao Te Ching*) and Chuang Tzu (the *Book of Chuang Tzu*), Taoism soon spread across China, offering ordinary people a way to personal salvation and healing. Taoists rarely found official acceptance but, as will be clear below, this was not a problem for them.

Confucianism – taking its name from K'ung Tzu, the teacher-scholar of the sixth century BCE – is sometimes described as a religion, sometimes not. Its origins are non-religious in that K'ung Tzu taught a moral and ethical way of life based upon his perception of good government. As indicated above, the Confucian world view ousted the older shamanic world view in terms of the social and political worlds of ancient China. K'ung Tzu's main concern with religion was that whatever was done should support the maintenance of order and control and thus be done properly; hence the Confucian distrust of spontaneity. Only gradually did Confucianism become religious through the elevation of K'ung Tzu to a god and the development of a state-religious apparatus, designed to honour hierarchy and control, and responding to a heavenly bureaucracy which exactly mirrored the bureaucracy of the empire.

The third great historic religion of China is Buddhism. Buddhism begins to emerge in China as the result of missionaries who arrive in the first century CE. By the end of the fifth century CE, Buddhism has become a major player in the religious and philosophical world of China. But it is not Buddhism as it was in India. China changed Buddhism as much as Buddhism changed China – perhaps more so. Chinese Buddhism draws upon Vedic materials, upon early Buddhist texts, but adds to this a specifically Chinese flavour. For example, the development of saviour deities – *bodhisattva*s – in Chinese Buddhism changes Buddhism from a faith in which you pursue a path of many rebirths in order to work off the effects of your *karma* to a situation in which, if you pray sincerely at death, the *bodhisattva* will come and rescue you and you need never be reborn again. One of the most famous developments of Buddhism in China, Chan (known as Zen in Japan), draws heavily upon Taoist ideas and practices.

Before going deeper into the foundations of Chinese religious outlooks, let me give a story which appears in the *Book of Lieh Tzu*. Scholarship is divided over the date of this text – ranging from third century BCE to third century CE. The fact is that it has been an important text from at least the fourth century CE and this story illustrates one of its teachings about the innate nature of things.

Mr T'ien was going to go on a long journey. To prepare for this, he sacrificed to the gods and held a grand feast to which he invited all his friends. When the dishes of fish and goose were brought in, Mr T'ien looked around benignly and said: 'How kind Heaven is to humanity. It provides the five grains and nourishes the fish and birds for us to enjoy and use.'

Everyone nodded in agreement, except for a twelve-year-old boy, the son of Mr Pao. He stood up and said: 'My Lord is wrong! All life is born in the same way that we are and we are all of the same kind. One species is not nobler than another; it is simply that the strongest and cleverest rule over the weaker and more stupid. Things eat each other and are then eaten, but they were not bred for this. To be sure, we take the things which we can eat and consume them, but you cannot claim that Heaven made them in the first place just for us to eat. After all, mosquitoes and gnats bite our skin, tigers and wolves eat our flesh. Does this mean Heaven originally created us for the sake of the mosquitoes, gnats, tigers and wolves?'

(Author's translation of the *Book of Lieh Tzu*: chapter 8)

The innateness of all life runs as a major theme throughout Chinese religious thought. It offers a key to understanding what Chinese religious thought sees as central to the meaning and significance of life. One interesting aspect of all three major faiths in China, Confucian, Taoist or Buddhist, is the lack of a creation story and the absence of any creator god or goddess. The origin of being is not divine in Chinese thought. It is natural. There certainly are myriad deities from deified generals to compassionate *bodhisattva*s. But there is no creator.

Instead, the nearest one gets to a creation theory is perhaps best captured in chapter 42 of the *Tao Te Ching*. Here, in what is in effect the core credal statement of those who believe in the Tao (Confucians as much as Taoists), is the explanation of the origin of life.

> The Tao
> gives birth to the One:
> The One
> gives birth to the two;
> The Two
> give birth to the three –
> The Three give birth to every living thing.
> All things are held in yin, and carry yang:
> And they are held together in the ch'i of teeming energy.
>
> (Author's translation of the *Book of the Tao Te Ching*: chapter 42)

The Tao is the origin and the ultimate natural force behind all natural forces. As such it comes close to the Buddhist concept of the Buddha nature which lies behind all Buddhas and in and through all worlds. From the Tao comes the One – the primal essence of being, the core of innateness. The One splits and forms the two – yin and yang. Yin and yang are total opposites. Yin is dark, female, cold, solid and wet. Yang is light, male, fiery, ethereal and dry. In other words, everything is either yin or yang. Yet each also carries the other within it. The classic symbol of Chinese religion is the yin/yang symbol in which both contain at their heart the seed of the other.

Yin and yang are complete opposites and the yin/yang symbol is not,

Figure 1: Yin and yang

as some in the West have claimed, a symbol of harmony. Instead it is a symbol of cosmic struggle as yin tries to overcome yang and yang seeks to overcome yin. It is from the dynamic energy thus generated that all life flows. It is important to stress here that yin and yang carry with them no sense of good and evil. There is no dualism in Chinese thought. Yin and yang are not better than each other: they simply are. One of the classic ways of describing yin and yang and their interaction is to use the seasons. Autumn and winter are yin. Yet at the very moment when yin triumphs – the heart of cold, dark winter – it also begins its inevitable decline and thus the yang forces begin to arise, leading to spring and summer – yang times. But again, when summer reaches its high point, the seasons begin to change and yin begins to arise again.

From the two come the three. This is conventionally seen to be the triad of Heaven (yang), Earth (yin) and humanity (both yin and yang). Heaven should not be thought of as a divine force. It carries overtones of this in later thinking and in popular thought. Heaven is where the celestial beings dwell, but classically Heaven is a moral force. It is the overarching principle of yang just as Earth is the undergirding principle of yin. Both are just natural forces or expressions of natural forces. Humanity's role is, however, strangely pivotal.

Carl Jung has said that humanity has to have a creation story which gives us a place and purpose in the cosmos, otherwise we will be crushed by the sheer awe-full-ness of the cosmos. In Chinese thought it is as the third member of the triad of Heaven, Earth and humanity that we find our meaning. It is the role of humanity to help maintain the balance between the conflicting and competing forces of yin and yang. This used to be symbolically expressed by the rituals which the Emperor, Son of Heaven, carried out. Each year at the summer solstice he would process to the Temple of Earth – yin – and make sacrifices to call the yin powers back and to begin to rebalance the world then in the grips of the height of yang. Conversely, at the winter solstice, he would process to the Temple of Heaven – yang – and make sacrifices to the yang powers to call them back.

At the same time, he would apologize and make symbolic and sacrificial atonement for the excesses of activity by humanity which threatened to tip the balance of the delicate line between yin and yang. Thus did the Emperor, symbol of humanity, play the role of humanity in the triad of relationships between Heaven, Earth and humanity.

Finally, in the credal statement from Chapter 42 of the *Tao Te Ching*, comes ch'i. Ch'i is the primal breath which brings all to life. Ch'i is energy, breath, vitality, semen, all that carries life within it. The flowing of ch'i is the realizing of the dynamic energy generated by the struggle between yin and yang. The nearest that Chinese mythology comes to a creation story relates to the first flowing of ch'i. The story goes that when the One spilt to produce the two – yin and yang – a being was created at the same time, or to be more exact, came into being. This being was called Pan Ku. Pan Ku found himself upon the barren, shapeless mass called Earth – yin. Here for untold millennia, he cut and carved, working away to create the landscape of the planet. He grew to unbelievable size. One day, worn out by his labour, he collapsed and died. It was only then that life came to the barrenness of the Earth. In classic creation story motifs, his flesh became the soil; his hair the trees and grass; his blood the waters; his breath the winds and rain. It took his ch'i, his energy, to bring life to the planet. His death released ch'i into the world and it has circulated ever since. An interesting side point in this story is the some-what humble origins ascribed to human beings. We, so the story goes, evolved from the parasites on his bottom!

Creation in Chinese religious thought is totally natural. The Pan Ku story is seen as exactly that, one attempt to clothe natural processes with a narrative. The importance of the story is the release of ch'i. Life exists so long as the ch'i exists. Once your store of ch'i, given to you at birth, has been exhausted, you die. In the pursuit of immortality, a core religious activity in China for at least 2,500 years, both the physical and the spiritual attempts to make the body survive for ever are rooted in ensuring that the ch'i remains active. Immortality in Chinese thought can only occur if you manage to preserve the physical body as the house of the ch'i. Once the body decays or the ch'i is exhausted, you die.

Thus to be in harmony or balance with nature, we need to be natural, to go with the flow of nature, to retain the breath of life, to be part of the great forces of existence. The big issue, of course, is how does nature flow?

For the Confucians, nature, the Tao, flowed through ordered and regular relationships within a clear hierarchy. For them, the Tao has always been more of a moral, ethical force than a pre-existing origin of the Origin. The *Tao Te Ching* became one of the key texts of Taoism precisely because it speaks of the Tao as ultimate ultimateness. For the Confucians, the Tao was more regulatory than creative. This is well captured in the following text taken from Confucius's *Analects*:

> K'ung Tzu [i.e. Confucius] said: 'Riches and honours are what people want. If these cannot be gained by the proper Tao, they should not be kept. Poverty and meanness are what people dislike. If it had been obtained by the right Tao, then I would not try to avoid them.'
>
> (Author's translation of the *Analects*: Book 4, verse 5)

This moral sense to the Tao, a moral force underpinning society, is central to Confucian thinking. In a later section of the *Analects*, K'ung Tzu makes the hierarchical nature of his understanding of the Tao very clear. For Confucians, the Tao, the natural order, was clearly hierarchical and patriarchal. At the top of the human hierarchy was the Emperor, guided by the Tao, by Heaven. If decisions were made by lesser mortals, those further down the hierarchical chain, then they were lesser commands and less moral and ethical. In K'ung Tzu's world, authority came firmly from the top. Anything less was less.

> K'ung Tzu said: 'When the Tao prevails in the world, the rites, music and punitive, military expeditions are initiated by the Emperor. When the Tao does not prevail in the world, they are initiated by the lesser lords. . . . When the Tao prevails in the world, policy is not in the hands of the Counsellors. When the Tao prevails in the world there is nothing for the ordinary people to argue about.'
>
> (*Analects*: Book 16, verse 2)

This text is close to a key text in the *Tao Te Ching* but the differences are instructive. The *Tao Te Ching* has often been described as a manual for leadership. But while Confucians place emperors at the top of their tree, the *Tao Te Ching* and Taoists place direct experience of the Tao, or insight mitigated by the sage, the recluse, who knows intuitively at the apex. Then and only then comes the emperor. The text says:

> The highest form of government
> Is what people hardly even realize is there.
>
> Next is that of the sage
> Who is seen, and loved, and respected.
>
> Next down is the dictatorship
> That thrives on oppression and terror –

And the last is that of those who lie
And end up despised and rejected.

The sage says little –
    and does not tie the people down;

And the people stay happy
Believing that what happens
                                    happens naturally.
                                    (*Tao Te Ching*: chapter 17)

The Confucian world view saw nature as orderly and that order was hierarchical and therefore that hierarchy was natural. This then manifested itself in what was at its best benevolent patriarchy. Women were subject to men; sons to fathers; wives to husbands; men to their lords; the lords to the Emperor; the Emperor to Heaven; and Heaven to the Tao. Disturb this order and society would break down, declared the Confucians. This is the origin in philosophical terms of the famous rigid Chinese social structure still visible today even under layers of Communist rhetoric. In this model, care of the rest of the natural world lay within the orbit of the structure of hierarchies. Any imbalance in nature's forces was taken as a sign that somewhere the structure of responsibilities had slipped. Earthquakes, floods, famines or any other such natural phenomena were interpreted as the displeasure of Heaven at the disruption of the hierarchy by wrongful human behaviour. It was for this that the Emperor went and offered sacrifices and apologies each year.

But unnatural events – comets in the sky, strange mutant births of animals – were even more dire. These were often taken as signs that Heaven had withdrawn the mandate to rule from the ruling family. When such thoughts as these were expressed, the ruling dynasty rarely lasted much longer. The Confucian world view was one of morality and punishment, ethics and rewards. You disturbed the natural balance at your peril. However, this did not mean you did nothing. Confucian thought allowed for many types of development so long as they did not disturb social order.

Taoism took a very different view of nature and humanity. It saw nature as spontaneous and as having within it a natural balance which did not need rigid structures; indeed, which broke out of such structures. The fourth-century BCE classic text, the *Book of Chuang Tzu*, puts it very bluntly in chapter 7 where two sages or aspiring sages are debating. One, Chien Wu, has recently visited another teacher and is now being questioned by the eccentric sage Chieh Yu. Chieh Yu asks Chien Wu what the teacher told him:

Chien Wu replied, 'He said to me that the nobleman who has authority over people should set a personal example by proper regulations, law and practices. The corollary of this will be that no one will disobey him and everyone will be transformed as a result.'

Eccentric Chieh Yu said, 'That would ruin Virtue. If someone tries to govern everything under Heaven in this way, it's like trying to stride through the seas or cut a tunnel through the river or make a mosquito carry a mountain. When a great sage is in command, he doesn't try to take control of externals. He first allows the people to do what comes naturally and he ensures that all things follow the way (Tao) their nature takes them.'

(Palmer with Breuilly 1996: 60–1)

The Taoists mock the attempts of humanity to control and channel the flow of nature. The hierarchies and structures of the Confucians make no sense to them and, worse, are seen as a perversion of the true Tao, the real way. This led the Taoists to seek their solace in remote places, as far away from the hierarchies and structures of the Confucian world as possible. The sage on the mountain, the man alone – itself quite an abhorrent notion to many Confucians – became the symbol of the Taoist way, as it still is today.

One very important function of Taoism with regard to nature is the role of cosmic liturgies. These liturgies are designed to rebalance the cosmos in much the same way as the Emperor rebalanced the cosmos in the strange mixture of shamanic and Confucian rituals at the solstices at the temples of Heaven and of Earth. In the Taoist liturgies, which are often paid for and commissioned by local communities who feel things around them are out of tune, out of balance, the Taoist priest enacts a ritual dance of the cosmos. In so doing he draws into the microcosm of himself the universe and then rebalances the cosmos as well as making amends for any actions of humanity which have disturbed nature. The following text from the Ling-pao Five Talismans liturgy gives some idea of what this involves:

The visible world was generated.
The workings of water and fire,
Life and death,
The myriad kalpas, and
The light of primordial yang were initiated.
The two principles of yin and yang
Used them to carve out the three realms.
The holy sages mounted them,
To attain union with the transcendent.
The five sacred peaks hold them,
And are thereby filled with spiritual power.
All things possessing them have life breath (ch'i).

(Saso 1978: 201)

Taoism today is recovering from the almost total destruction which descended upon it during the Cultural Revolution in China (1966–75). In 1985 there were just 2,500 ordained Taoist monks and nuns. Today this number has risen to around 13,500. The Taoists are also the only ancient religious community in China to have made a formal commitment to environmental work. In 1995 they issued their *Statement on Taoism and the Environment* and formally joined the Worldwide Fund for Nature's programme on religion and conservation. At the same time, they launched a major environmental initiative on the sacred mountains of China, as is discussed below.

The following is taken from their statement:

With the deepening world environmental crisis, more and more people have come to realize that the problem of the environment is not only brought about by modern industry and technology, but it has a deep connection with people's world outlook, with their sense of value and with the way they structure knowledge. Some people's ways of think-ing have, in certain ways, unbalanced the harmonious relationship between human beings and nature, and overstressed the power and influence of the human will. People think that nature can be rapaciously exploited.

This philosophy is the ideological root of the current serious environ-mental and ecological crisis. On the one hand, it brings about high productivity; on the other hand, it brings about an exaggerated sense of one's own importance. Confronted with the destruction of the Earth, we have to conduct a thorough self-examination on this way of thinking.

We believe that Taoism has teachings which can be used to counteract the shortcomings of currently prevailing values. Taoism looks upon humanity as the most intelligent and creative entity in the universe (which is seen as encompassing humanity, Heaven, Earth within the Tao).

There are four main principles which should guide the relationship between humanity and nature:

1 In the *Tao Te Ching*, the basic classic of Taoism, there is this verse: 'Humanity follows the Earth, the Earth follows Heaven, Heaven follows the Tao, and the Tao follows what is natural.' This means that the whole of humanity should attach great importance to the Earth and should obey its rule of movement. The Earth has to respect the changes of Heaven, and Heaven must abide by the Tao. And the Tao follows the natural course of development of every-thing. So we can see that what human beings can do with nature is to help everything grow according to its own way. We should culti-

vate in people's minds the way of no-action in relation to nature, and let nature be itself.

2 In Taoism everything is composed of two opposite forces known as yin and yang. Yin represents the female, the cold, the soft and so forth; yang represents the male, the hot, the hard and so on. The two forces are in constant struggle within everything. When they reach balance, the energy of life is created. From this we can see how important harmony is to nature. Someone who understands this point will see and act intelligently. Otherwise, people will probably violate the law of nature and destroy the harmony of nature.

There are generally two kinds of attitude towards the treatment of nature, as is said in another classic of Taoism, *Bao Pu Zi* (written in the fourth century BCE). One attitude is to make full use of nature, the other is to observe and follow nature's way. Those who have only a superficial understanding of the relationship between humanity and nature will recklessly exploit nature. Those who have a deep understanding of the relationship will treat nature well and learn from it. For example, some Taoists have studied the way of the crane and the turtle, and have imitated their methods of exercise to build up their own constitutions. It is obvious that in the long run, the excessive use of nature will bring about disaster, even the extinction of humanity.

3 People should take into full consideration the limits of nature's sustaining power, so that when they pursue their own development, they have a correct standard of success. If anything runs counter to the harmony and balance of nature, even if it is of great immediate interest and profit, people should restrain themselves from doing it, so as to prevent nature's punishment. Furthermore, insatiable human desire will lead to the over-exploitation of natural resources. So people should remember that to be too successful is to be on the path to defeat.

4 Taoism has a unique sense of value in that it judges affluence by the number of different species. If all things in the universe grow well, then a society is a community of affluence. If not, this kingdom is on the decline. This view encourages both government and people to take good care of nature. This thought is a very special contribution by Taoism to the conservation of nature.

(China Taoist Association 1995: 3–4)

This extract gives a powerfully clear vision of what Chinese folk religion in general, but Taoism in particular, has to say about the environment.

Of the two other great religions of China, one has ceased to function in any significant way: the Confucian temples are all empty. While even those Buddhist and Taoist temples which are now merely museums will have little offerings placed before the statues and each faith has thousands of functioning temples, no-one lays offerings before the statues in the great Confucian centres. The religion of Confucianism, at least in China, is dead. Even outside China, it has little appeal. The empire has gone and with it a whole way of life which supported Confucianism. Many business commentators speak glibly about the Confucian ethic which it is believed underpins the rise of countries such as China, Korea, Taiwan and Singapore. Confucian social structures certainly still exist and shape Chinese attitudes to society and to life. But as a religion it is gone.

Buddhism, meanwhile, is experiencing a revival. Alongside Christianity, it is the rising religion of modern China. It has powerful teachings about care for nature, which are expressed most forcefully through the compassion of the *bodhisattvas*. The *bodhisattva* is one who through countless lives has gained such good merit as to be free to leave the world of birth, death and rebirth. They could pass over into nirvana. But, moved by compassion, they hold back. Using their vast store of good deeds, they ransom souls from the ten hells and from the horrors of rebirth. All those who genuinely believe and genuinely pray with sincerity will be saved. The compassion of the *bodhisattvas* is one of the joys of Chinese Buddhism, offering millions a vision of hope of escape from the sufferings of existence. It is in the compassion of the *bodhisattvas* that Chinese Buddhism finds its clearest advocacy of care for nature.

Of all the *bodhisattvas* in Chinese Buddhism, the most popular is Kuan Yin, the goddess of mercy. The following is a description of her saving powers and her compassion for all life forms:

> She delivers from the eight terrors,
> Saves all living beings,
> For boundless is her compassion.
> She resides on T'ai Shan,
> She dwells in the Southern Ocean.
> She saves all the suffering when their cries reach her,
> She never fails to answer their prayers,
> Eternally divine and wonderful.
>
> (Palmer, Ramsay with Kwok 1995: 91)

This compassion is a hallmark of Buddhism in China. Through what one might describe as passive compassion, the lands around Buddhist temples and monasteries have often become *de facto* sanctuaries for wildlife, because religious tradition forbade killing on or near such places. However, Chinese Buddhism has yet to make the step other Buddhist traditions have made of turning a passive compassion into an active one. It is to be hoped that this will begin to happen as religious life in China is allowed to be restored more fully.

Apart from the three major faiths, there is what is variously called traditional or folk Chinese religion. This is, in fact, the matrix within which most Chinese conduct their religious life. It is a mixture of shamanic, Taoist, Buddhist, Confucian and now Christian and even Maoist ideas, images and practices. A worshipper might wear a lucky Mao badge, rub a crucifix for good fortune while offering incense to the ancestors and going to a fortune teller who calls up the spirit of the Buddha of the future to give a reading on a chart marked with the yin/yang symbol.

Most Chinese find it hard to specify exactly which religion they belong to for they take and use parts of many. It is in this folk-religion context that one of Chinese religion's most significant contributions to environmental awareness arises. This is the art of feng shui – geomancy. 'Feng shui' means literally wind–water and refers to the need to position any building, tomb or even your bed according to the prevailing natural forces in the area. Feng shui has helped shape the characteristic Chinese landscape where buildings complement the natural features, rather than attempt to overawe them; where trees and bushes are planted to help buildings merge into the landscape; where natural materials predominate over humanly constructed materials; where height and breadth of buildings are strictly controlled by the height and breadth of natural features around them. In other words, it is a way of building and relating to the landscape which sees the existing landscape as full of powers and forces, meaning and purpose long before humanity came to contribute.

Feng shui is still practised widely in China and elsewhere by Chinese communities. Its often commonsense approach to building and its emphasis on general suitability of certain styles to certain landscapes are increasingly being recognized in the West as valuable in terms of understanding the interplay between the natural and humanly constructed environments. In China, sadly, official Communist policy has ignored these ancient and tried and tested methods while the new 'grow rich quick' entrepreneurs just give it a nod but seem rarely to allow it to affect what and where they build.

But in those circumstances where religions or where families are building for spiritual purposes, the feng shui outlook is being brought back again. In feng shui, Chinese religious and philosophical theories find practical expression in relationship to the everyday lives of ordinary people. It is one way in which Chinese folk religion helps build a more Tao-like way of life.

Finally, I want to look at the sacred landscape of China itself. China is still an inherently sacred landscape (see Palmer 1996). You cannot travel more than a few score miles in China without coming to a sacred mountain or sacred river. In particular the scores of sacred mountains across China offer islands of environmental protection which have almost all vanished elsewhere. I would say that there are over a hundred such sacred mountains, often only known to be sacred in their immediate environment but protected from exploitation by the devotion of local families and villages – folk religion again. But among these are the nine formal sacred mountains of China. Four of these are Buddhist and five are Taoist. The Taoist ones stretch back into the shamanic mists of early China. The Buddhist ones have been developed in the last thousand to fifteen hundred years.

It is interesting to note the very different experiences you have on a Taoist sacred mountain from those on a Buddhist one. On Buddhist mountains, such as Emei Shan in Sichuan Province or Wutai Shan in Shanxi Province, the path is simply the most effective way to travel from point A to point B – usually from great monastery to great monastery. The guidebooks of old stressed where to stop for scenic views and for contemplation of nature. On a Buddhist mountain you are an observer.

On Taoist mountains, rooted far, far back in indigenous Chinese culture, the situation is completely different. On the Taoist mountains, such as the greatest of all sacred mountains, Tai Shan in Shandong or Hua Shan in Shanxi Province, the path is the way, is the Tao. By walking the path, you enter into Taoism and become part of the flow of Tao. The path is the purpose, not the monasteries or temples en route. At every bend, every interestingly shaped rock, every trickling stream and every unusual tree you stop to hear the story of how it came to be thus. You narrate your way up the mountain. On these paths, there are few places where you look at the view, for the simple reason that you are no longer the observer. You are the observed. You are part of the phenomena of the mountain.

No clearer example can be given to help us enter into the way in which Taoism incarnates its belief in the vital life of all nature of which we are but a part. This is why the China Taoist Association has launched, as its first major environmental protection programme, a scheme to help pre- serve the natural and spiritual beauty of the sacred mountains. This project, undertaken in collaboration with the Alliance of Religions and Conservation, is currently surveying the Taoist sacred mountains and developing religiously founded conservation programmes. From this will flow protection of the mountains, educational projects for the visitors and training programmes for Taoists. From the sacred mountains it is intended that a sense of the sacred will begin to travel back down to the villages and cities of China.

Perhaps the extent to which this spirit is reinvigorating Taoism can be seen in the example of Zhongyue Temple on Song Shan, Henan Province. Here, in what is in effect a shamanic temple taken over but virtually unchanged by Taoists, stand seventy-two statues. For many years they have been neglected. In no other temple in China do these seventy-two deities exist. Today they are freshly restored and painted and much revered. The reason is simple. They are the seventy-two judges who punish the wicked for crimes against the Tao. And these crimes are listed on the boards they hold. One punishes those who pollute clean water; another punishes those who kill animals for pleasure not for food; who trap birds; who cut down ancient trees; or destroy habitats for no good reason.

The relevance of these statues and their return to favour is symptomatic of the awakening realization in Chinese religion of the fact that there lies within all the traditions, but perhaps especially in Taoism and folk religion, a wisdom about how to live in balance with the rest of nature and how to become again aware of the flow of the Tao, the flow of nature of which we are but a part.

3 Religion and nature

The Abrahamic faiths' concepts of creation

His All-Holiness Bartolomeus, Professor Rabbi Arthur Hertzberg, and Fazlun Khalid
(Compiled and edited by Martin Palmer)

The world of contemporary ecology is the world of crusading movements seeking salvation for the earth's ecosystems. As such its roots lie in one direction in the Abrahamic world views and one can argue that the environmental movement is but the latest expression of this particular way of understanding the world. Indeed, it bears all the hallmarks of a missionary venture and the origin of the environmental movement in the West is not pure accident.

However, there is a fundamental difference between the Abrahamic faiths and much of contemporary ecology. Ecology as espoused by bodies such as the United Nations, the World Bank, multinationals and many quasi-scientific bodies such as the Worldwide Fund for Nature or the International Union for the Conservation of Nature and Natural Resources focuses on the assumption that creation is here for human use and is of significance primarily, if not entirely, from a functional point of view. It can be said that the implicit logic behind the protection of the environment is expressed purely as a logic of convenience.

Within the context of this particular logic, the protection of the environment is understood as essential to human survival. It is, however, primarily valued for its usefulness. The environmental crisis is restricted to how the natural environment is used. The origin or cause of the natural environment is apparently of no concern, nor do most environmentalists search for an explanatory 'meaning' of cosmic good, orderliness, harmony, wisdom or the beauty of nature. It is considered possible that the material stuff of nature was created by unknown 'powers from above', or possibly by 'chance'. In any case, in much of conventional environmental thinking, it is not the interpretation of cause and aim which gives *meaning* to matter in the ecologist's world view. It is the usefulness of the material world which gives it meaning to the ecologist.

Stemming from this logic of convenience, the ecological movements today demand that rules be set down for how humanity should use nature. Ecology aspires to be a practical ethic controlling and shaping human behaviour towards the natural environment. But ecology as an ethic raises the questions: who determines the rules of human behaviour and by what authority? what logic makes these rules compulsory? and what is the source of their validity?

The necessity for ecological ethics, and, one could argue, their correctness, is borne witness to by their evident usefulness. It is logical that in order for humanity to survive on the earth, a new way of living with the natural environment that enables human survival must be found.

However, the rationale which has fuelled the protection of the natural environment as something purely instrumental is inherently caught up in the very values and beliefs about nature as commodity which have themselves led to the destruction of the natural environment. Humanity does not destroy the environment because of motives of irrational self-gratification. Rather, it destroys it by trying to take advantage of nature in order to secure more conveniences and comforts in daily life.

The logic of the destruction of the environment is precisely the same as that of the protection of the environment. Both 'logics' confront nature as an exclusively useful given. They do not give it any other *meaning*. Ecologists demand limited and controlled exploitation of the natural environment – that is, a quantitative reduction – which would also permit its further long-term exploitation. They ask for the rational limitation of use – in other words a new kind of consumeristic rationalism – which is 'more correct' than the consumeristic rationalism of current patterns of exploitation. They ask for consumeristic 'temperance' in consumerism.

This attempt, even though it appears to be extremely rational, is, by definition and in practice, irrational. By definition it is contradictory because consumerism cannot come into conflict with consumerism. And it is irrational in practice because the majority of the earth's population will not accept being deprived of the convenience and comforts which the destruction of the environment has so far secured for a minority of 'civilized' societies.

The Abrahamic religious traditions of Judaism, Christianity and Islam preserve the attitude that the meaning and significance of nature – creation – is not exclusively instrumental. In these traditions, the world is a creation of God. The use of the world by humans constitutes a pragmatic relationship between humanity and God, because God gives and humanity receives the riches of nature as an offering of God's divine love for the sake of the whole world. It is this outlook that will be explored in this chapter.

Perhaps it is helpful here to define the term and the origin of the phrase 'Abrahamic faiths'. The three great faiths of Judaism, Christianity and Islam are obviously linked by common sources, concepts and traditions, though each interprets them differently. In the past the term 'People of the Book' was used. However, this begged the question of which book – Torah, New Testament or Qur'an? It also carries with it the assumption that other religions are without significant sacred and revelatory literature.

What all three religions do hold in common is the importance of the Patriarch Abraham. Judaism, of course, sees Abraham as the central patriarch in the narrative of Israel as recorded in Genesis within the Torah. Christianity, likewise, holds Abraham in high regard, seeing his relationship with God and trust in God as a forerunner of Christ. In Islam Abraham is held in the highest regard as the patriarch from whom the Arabic peoples descended, as well as the model of a faithful servant of God. Hence the use of this term to describe the three linked faiths of Judaism, Christianity and Islam.

The Torah, the most holy book of Judaism, which also forms the first part of what Christians term the Old Testament, has this to say concerning Abraham:

> The Lord said to Abram, 'Leave your country, your family and your father's house, for the land I will show you. I will make you a great nation; I will bless you and make your name so famous that it will be used as a blessing.'
>
> (Genesis 12: 1–2)

Judaism sees Abraham as the founder of not just its faith but of its people. The covenant formed with Abraham was the first covenant that God created, giving rise to the term for Jews as 'the people of the covenant'.

But the covenant also established a crucial ethical point, a point of justice. It enabled Abraham to demand justice of God. So, when Abraham stood before God after God had declared that he was about to destroy the sinful city of Sodom, Abraham demanded of God that he give moral justification for this act, saying 'Shall not the judge of all the earth do justice?' (Genesis 18: 25).

It is perhaps worth noting that the Torah refers to 'Abram' and then to 'Abraham'. In Hebrew the significance of this is clear. 'Abrah' means father of one people; 'Abraham' means father of many peoples.

In the Christian New Testament, the link of Jesus to Abraham is made explicit in both the genealogies of Christ in the Gospels of Matthew and Luke, and in the vision of why Christ took human flesh and dwelt among us. When Mary is told she will bear the Messiah, the Son of God, she sings the Magnificat, her song of praise for the gift of Jesus. In the Magnificat, she describes how God will pull down the mighty from their seats and exalt the humble and meek; how he will fill the hungry with good things and send the rich away empty. She ends her song of praise with: 'He has come to the help of Israel his servant, mindful of his mercy – according to the promise he made to our ancestors – of his mercy to Abraham and to his descendants for ever' (Luke 1: 54–5).

In the Qur'an, where Abraham is known as Ibrahim, there is an entire *sura* or section named after him. As the founder of the Kabba at Makkah, he is held as being the first true believer and as a model for all believers. As sura 16, verses 120–2 says:

> Ibrahim was indeed a model,
> devoutly obedient to God,
> True in faith,
> he did not honour any gods but God.
>
> He showed his gratitude
> For the favours of God,
> Who had chosen him and guided him
> On the Straight Path.

Thus, to understand the Abrahamic faiths of Judaism, Christianity and Islam, we need to understand what it was that Abraham believed and what was revealed to him. In doing so we can begin to see how these beliefs affect the attitudes and actions of followers of these three faiths towards the natural world.

Central to the Abrahamic understanding is belief in One God, creator and sustainer of all that has been, is and will be. Nothing exists but for the Will of God. As the Book of Genesis in the Torah says: 'In the beginning God created the heavens and the earth. Now the earth was a formless void, there was darkness over the deep, and God's spirit hovered over the water' (Genesis 1: 1–2).

The Qur'an says:

> To Him is due
> the primal origin
> of the heavens and the earth:
> When he decrees a matter
> He says to it, 'Be',
> And it is.
>
> (Sura 2: 117)

Thus all creation is of God and has meaning and purpose before God.

A moving story is told of the Chief Rabbi of Jaffa, Abraham Isaac Kook, in the early 1900s. A visitor, Rabbi Aryeh Levine, went to stay with him:

> He received me warmly ... and after the afternoon prayer I accompanied him as he went out into the fields, as was his wont, to concentrate his thoughts. As we walked I plucked some flower or plant; he trembled, and quietly told me that he always took great care not to pluck, unless it were for some benefit, anything that could grow for there was no plant below that did not have its guardian above. Everything that grew said something, every stone whispered some secret, all creation sang.
>
> (Levine 1961: 15–16)

In Christianity St Francis expressed the vision of all creation being part of one family – for he taught his followers to consider the animals and birds as their brothers and sisters. This was not some silly sentimentality, but a profound model of our relationship, under God, to the rest of creation. It is captured in these lines from his Canticle of the Creatures:

> Be praised then, my Lord God,
> In and through all your creatures,
> Especially among them,
> Through our Noble Brother Sun. . . .
>
> Be praised, my Lord, through Sister Moon and all the stars;
> You have made the sky shine in their lovely light.
>
> In Brother Wind be praised, my Lord,
> And in the air,
> In clouds, in calm,
> In all weather moods that cherish life. . . .
>
> Through our dear Mother Earth be praised, my Lord.
>
> (Cited in Reidy 1986: 37)

Within the Orthodox Christian faith, Fyodor Mikhail Dostoevsky, the great nineteenth-century Russian novelist, said:

> Love all God's creation, the whole of it and every grain of sand. Love every leaf, every ray of God's light. Love the animals, love the plants, love everything. If you love everything, you will perceive the divine mystery in things. . . . Love the animals: God has given them the rudiments of thought and untroubled joy. Do not, therefore, trouble them, do not torture them, do not deprive them of their joy. Do not go against God's intent.
>
> (Cited in Dimitros 1990: 8)

In Islam, the Qur'an teaches that all animals live in community and that they are known and accountable before God:

> There is not an animal that lives on the earth, nor a being that flies on its wings, but forms part of communities like you. Nothing have We omitted from the Book, and they all shall be gathered to their Lord at the end.
>
> (Sura 6: 38)

It also denounces the arrogance of those who treat the rest of creation without respect:

> Do you not see that it is God whose praises are celebrated by all beings in the heavens and on earth, even by the birds in their flocks? Each creature knows its prayer and psalm – and so does God know

what they are doing. And yet, you understand not how they declare
His Glory.

(Sura 24: 41)

Within these interpretations of all creation arising from a loving creator
God, the questions have to be asked: 'What of humanity?' 'What role do
the Abrahamic faiths ascribe to human beings?' The answers in Judaism
are clear and the consequences are immense.

Psalm 8 addresses these questions directly:

I look up at your heavens, made by your fingers,
at the moon and the stars you set in place –
ah, what is man that you should spare a thought for him,
the son of man that you should care for him?

Yet you have made him little less than an angel.
You have crowned him with your glory and splendour,
made him lord over the work of your hands,
set all things under his feet.

sheep and oxen, all these,
yes, wild animals too,
birds in the air, fish in the sea
travelling the paths of the ocean.

(Verses 3–8)

For Judaism, humanity is the greatest of all God's creations yet seems
to be worthy of so little. The special significance of humanity is shown
in both versions of the creation story in Genesis, chapters 1 and 2. In
chapter 1 God specifically blesses humanity – unlike any other form of
creation. In the second account in chapter 2, God breathes into the dust
from which Adam is made and it is God's breath which brings life to
Adam.

Judaism sees humanity as little less than the angels, yet recognizes
human nature to be often indistinguishable from that of the beasts of the
field. It is this tension between potential and reality which acts as
the dynamic narrative of the Jewish scriptures and tradition.

Christianity inherited this tradition, as Jesus shows when he compares
the value of human beings to sparrows. Both are loved and cared for by
God, but human beings are a hundred times more important (Luke 12:
6–7). But the tension between potential and actuality is brought to a head
in the person of Christ and the reaction of those around him. Here is
humanity at its most divine. Yet most cannot accept him and bring him
to his death rather than be confronted by his reality.

All three religions see humanity as central to the loving purpose of
God's creation, even if we do not deserve it in our terms and within our
limited understanding. There is no concept here of humanity being just

one species among others, and it is important to understand why this is so.

The roots of the creation stories in the Torah, hinted at in the New Testament and expressed within the Qur'an are the rejection of earlier models of creation stories from the ancient Middle East. Until the composition of Genesis, the basic assumption of Middle Eastern creation stories was that the gods created out of self-interest and for their own amusement. Humanity was but a plaything of the gods, pawns on a chessboard. One has only to think of how Homer recounts the story of the Trojan Wars as essentially a struggle between conflicting parties of gods and goddesses who move human beings around in order to fulfil the whims or desire for revenge.

The radical shift which the writer of Genesis records is that this old model is discarded. Humanity, indeed all creation, is no longer just a backcloth against which the drama of the gods is enacted. Instead, creation becomes the loving expression of a purposeful creator God and humanity moves from being a plaything of the gods to the beloved special creation of God – almost the partner of God.

Two fundamental consequences emerge from attitudes such as these. First, that the world is not meant to be used by humans for their own purpose, but is the means whereby humans come into relationship with God. If humans change this use into egocentric, greedy exploitation, into oppression and the destruction of nature, then their own vital relationship with God is denied and refuted; a relationship that is predestined to continue into eternity. The second consequence is that the world as a creation of God ceases to be a neutral object for humanity's use. The world incarnates the word of the Creator, just as every work of art incarnates the world of the artist. The objects of natural reality bear the seal of the wisdom and love of their Creator. They are the word of God calling humanity to come into *dialogue* with God.

This revelation of the central meaning of humanity fundamentally changed how human beings saw themselves and how they valued themselves. From this, we argue, has sprung much of the sense of purpose and significance of human life which has fuelled the development, both spiritual and material, of human civilization.

As stated above, Judaism recognizes an ambiguity or tension in its teachings and understanding of our place within God's creation.

This tension is to a considerable degree shared by Christianity. In essence, at the heart of the biblical tradition, set within the very context of the story of creation itself, there is the dilemma of the relationship between power and stewardship over nature. As the Jewish Declaration on Nature, issued at the Assisi meetings in 1986, put it:

The world was created because God willed it, but why did He will it? Judaism has maintained in all its versions that this world is the arena that God created for humanity, half beast and half angel, to prove that it can behave as a moral being.

(Hertzberg 1986: 30)

Further to this is the point that the Talmud makes when it pictures humanity as being God's partner in the act of creation. Humanity is charged with not only being responsible for the earth, but with helping to shape it further, so that there is order and decency rather than chaos.

The belief that morality underpins the natural world is borne out in the teaching that God is also bound by this morality. As noted earlier, Abraham is able to argue and debate with God over the fate of Sodom, recognizing humanity's sin but also pleading for its possibilities of goodness. Abraham can assume that the God who demands justice will always act justly. In a similar way, argues Judaism, humanity has also been given dominion over nature but always within the context of morality, justice and compassion. We live in a tension between the scale of our power and the limits imposed by morality and conscience.

In recent years, not least among some within the conservation and ecology movements, it has become fashionable to lament the biblical emphasis on how much control humanity has over the rest of creation. Indeed, some have even traced the environmental crisis to the teachings in the Bible about having dominion and power. But the brutal fact is that this power exists. Whether this arose because those who read the Bible felt legitimated to take such power, or whether the Bible simply reflects the truth of the situation, is another debate. What we have to confront is the reality of power in our hands. Bemoaning this and wishing we could give it up and sink into relative obscurity as just one species among others will not do. We have this power. This is what the Abrahamic faiths confront with almost brutal honesty. The question is not whether we should have such power – we do – but what we do with it.

The response of Islam is to place such power within the confines and ethics of a greater power – that of God. Islam expresses this in the notion of humans being khalifas. A khalifa is a vice-regent, someone appointed by the Supreme ruler to have responsibility over a given area in an empire. The Qur'an, in sura 2: 30, uses this phrase as a description of the role of humanity.

Behold Thy Lord said to the angels: 'I will create a vice-regent on earth.' They said, 'Will Thou place there one who will make mischief and shed blood? We celebrate Thy praises and glorify Thy holy name.' He said, 'I know what you know not.'

We have been given great power and authority by God, but only to be used on God's behalf, not for our own ends and ambitions. The Qur'an makes it clear that any abuse of this power, any wanton or wasteful use of the world's natural resources, is repugnant to God and thus to Islam. The Islamic Declaration on Nature, issued at Assisi in 1986, says the following:

> For the Muslim, humanity's role on earth is that of a khalifa, vice-regent or trustee of God. We are God's stewards and agents on Earth. We are not masters of this Earth; it does not belong to us to do what we wish. It belongs to God and He has entrusted us with its safe-keeping. Our function as vice-regents, khalifas of God, is only to oversee the trust. The khalifa is answerable for his/her actions, for the way in which he/she uses or abuses the trust of God.
>
> (Naseef 1986: 23)

Christianity contains within it the tension model of Judaism, the stewardship model of Islam and yet other responses. For example, some would see Christianity as offering a dominion model which arises from the model of a dominant Lord God, a figure of authority and power. Others would see a stewardship model, similar to Islam, where we have to take care of what is only lent to us for a limited time. Yet another view walks neither of these paths but offers a very different perspective from the Orthodox Christian tradition.

The Christian faith teaches that Christ is part of the creating Trinity of Father, Son and Holy Spirit. Thus we can describe Christ as Creator in the way that St John does in the prologue to his Gospel:

> In the beginning was the Word:
> The Word was with God
> and the Word was God.
> He was with God in the beginning.
> Through him all things came to be
> not one thing had its being but through him.
>
> (John 1: 1–3)

St Paul tells us that the Creator emptied himself of his power and came to earth as a child – a weak and defenceless child. He took upon himself the role of a servant. So at the heart of Christianity is the Creator made created, the Mighty Lord become a child. The Master of all became the servant of all.

And it is this model which informs, infuses and inspires the Orthodox understanding of our place in creation. Certainly, we are to take seriously the power we have, the power of life and death over so much of creation, human, animal, plant, fish or fowl, over habitats, forests, lakes, rivers, and the good earth. And our response? It is to follow the example of Jesus Christ and to turn this power away from mastership, dominion and destruction towards servanthood. We should serve all creation in the same way that a priest in the Orthodox tradition serves the community and serves God.

The heart of the Orthodox Christian faith is the Eucharist – meaning thanksgiving. Here is celebrated the offering of bread and wine, just as Jesus commanded at his Last Supper with his disciples. At the Eucharist the community offers the work of its hands, the fruit of the earth, to God. It is received back sanctified and blessed as spiritual food. In the document *Orthodoxy and the Ecological Crisis*, published by the Ecumenical Patriarch of Constantinople in 1990, the following is written:

> Just as the priest at the Eucharist offers the fulness of creation and receives it back as the blessing of Grace in the form of the consecrated bread and wine, to share with others, so we must be the channel through which God's grace and deliverance are shared with all creation. The human being is simply yet gloriously the means for the expression of creation in its fulness and the coming of God's deliverance for all creation.
>
> (Dimitrios 1990: 8)

Put simply, we are called to be a means of blessing to creation. Look about you. Do you see much in our world today that speaks of our being a blessing to the rest of creation? Far from it. We see species dying out; peoples dying out; habitats and ecosystems being destroyed. Why?

All three of the Abrahamic faiths would give a similar answer – human wickedness or sin. For just as we have the choice of following the revealed way and Will of God, so conversely we have the ability to choose to disobey, to rebel and thus to sin against God and against God's creation.

Perhaps the most disturbing aspect of the teaching of Judaism, Christianity and Islam is the belief that we can choose to disobey. Our freedom under God is immense. Our capacity for good is vast. Our capacity for evil is likewise vast. And often this evil comes about because of foolishness, greed, pride or ignorance. Yet the end result is the same – destruction.

Thus all three faiths, whichever model they present – dominion under the guidance of the Lord, dialogue between power and stewardship or mighty ones becoming servants – call for us to repent of what has gone wrong and offer us forgiveness and the chance to start again. The Qur'an puts it vividly: 'Corruption has appeared over land and water on account of what humanity's hands have wrought' (sura 30: 41).

The prophet Isaiah put it starkly:

> Ravaged, ravaged the earth,
> despoiled, despoiled,
> as the Lord has said.
> The earth is mourning, withering,
> the world is pining, withering,

the heavens are pining away with the earth.
The earth is defiled
under its inhabitants' feet,
for they have transgressed the law, violated the precept,
broken the everlasting covenant.

(Isaiah 24: 3–6)

In all three Abrahamic faiths, the tension between the ability to do the Will of God and the ability to rebel is to be found. And alongside it is the ability to return to the right path, to follow again the Path of God revealed in the Scriptures. This is why Jews, Christians and Muslims can look upon the present state of the world with comprehension, distress and hope; comprehension because we can see what happens when people forget that this world is not ours but God's; distress when we see the consequences of evil doing, of pride, foolishness and greed which go against the teachings of God; but most of all with hope, for we all believe that if we turn again to God, He will meet us and bring us home to Him. By returning to God we can start again. The past can be forgiven and we can hope that things can be fundamentally different.

More than this, we can find within our three traditions paths, models and understandings which can lead us towards a new way of living in relationship not just with God, but with all that God has loved and created – the whole of creation. Only when we confront matter and all of nature as the work of a person-Creator do we establish a true *relationship* with and not a domination over nature. Only then will it be possible to speak about an 'ecological ethic' which does not take its regulatory character from conventional rational rules, but arises out of the need for a person to love and be loved within the context of a personal relationship. The reason there is beauty in creation, so that it is an object of love, is that it is a call from God to humanity; it is a call into a personal relationship and to communion of life with God; a relation which is vital and life-giving. Contemporary ecology could then be the practical response of human beings to God's call; the practical participation in our relationship with God.

If there is a future for the ecological demands of our contemporary times, this future is based, we believe, in the free encounter of the historical experience of the Living God with the empirical confirmation of God's active word in nature.

The challenge that lies before us as people of the three Abrahamic faiths is simple: to live out in our lives, in our institutions and in our witness to the rest of the world the reality of being followers, believers in the God of Creation; to do so wherever possible in cooperation between the three Abrahamic faiths; also to seek points of contact, areas of concern and issues that unite us with those faiths which come to the issue of the environment from different perspectives. For we need as faiths to listen and learn in order to know what it is that we have to offer and share.

The challenge is simple. Its consequences are enormous, life-changing and, as yet, far from being fulfilled.

4 Pantheism

Stephen R. L. Clark

PHILOSOPHICAL PANTHEISM

Arnold Toynbee, conversing with Daisatsu Ikeda in the early 1970s, declared 'a right religion is one that teaches respect for the dignity and sanctity of all nature. The wrong religion is one that licenses the indulgence of human greed at the expense of non-human nature. I conclude that the religion we need to embrace now is pantheism, as exemplified in Shinto, and that the religion we now need to discard is Judaic monotheism and the post-Christian non-theistic faith in scientific progress, which has inherited from Christianity the belief that mankind is morally entitled to exploit the rest of the universe for the indulgence of human greed' (Toynbee and Ikeda 1976: 324). Other dedicated environmentalists, 'deep ecologists', self-styled 'pagans' and even Christian theologians have also preferred pantheism (see Wood 1985; Martin 1993). What theologians and environmentalists might mean by 'pantheism' remains obscure. 'It is almost as though literary commentators [and others] have entered a kind of silent conspiracy never to challenge one another as to the exact meaning of these ideas, or to the appropriateness of their invocation' (McFarland 1969: 127).

The term itself (though not, of course, the things or theories it identifies) dates only from John Toland in 1705,[1] and was long used, like 'Spinozism' or 'atheism', as a term of theological abuse. But pantheists distinguish themselves from theists and from atheists alike: typically, they take no part in conventionally religious rituals (though they may invent or rediscover their own seasonal events) and doubt that the world is run by ordinarily moral notions, but still affirm that something well worth worshipping (whatever quite that means) is to be found 'in nature'. Some pantheists mean by their 'pantheism' that every single thing is divine; others (whom I shall be reckoning 'philosophical pantheists') that it is the Whole, the unitary All, that is the only God. Some may believe that this Divine is best appreciated in the human self, that 'we' are the place where God and Nature becomes self-aware. Others may affirm instead that God and Nature is the Whole of which we ourselves are only and forever particles, that God is to be acknowledged (but not fully known) in a Nature of which human knowledge is not, after all, a model. Most will deny that 'God' is any sort of person, or that It is moved by any 'personal' intention or love-liking. God's 'omnipotence' rests in Nature's power to do anything that is possible, God's omniscience in the simple fact that the Truth contains all truths, God's love only in the persistence of the All in being. Historically, most pantheists (or most of those whom we would reckon pantheists) would deny that the world was going any-where it had not been before. There might be cycles of growth and decay, local or even universal: there would be no new conclusions. More recently, some of those now reckoned pantheists have imagined some long process to a final End: the point at which all stories are complete, all separate (or seemingly separate) entities are brought together. That image, of the Omega Point or the Conflagration, was a Stoic theme – but Stoics expected that that Age or Instant would be succeeded, once again, by the self-same multiplicity that had preceded it. Most pantheists have also been determinists; a few perhaps believe that God – and therefore the world – is free and undetermined, that the future is not contained in any past at all.

These permutations mean that Toynbee's 'pantheism' may be quite unlike the doctrines commonly associated with such well-known panthe-ists as the Enlightenment philosopher Benedict Spinoza, or Greek philosophers such as Chrysippus or Marcus Aurelius. Such philosophical pantheists assert that it is the All, the unitary Totality, which is God. The claim is a rational one: if 'God' means that than which none greater can be conceived, it must follow that God contains everything of any value. If even the least valuable thing lay outside God, there would be something greater than God alone – the sum of God and that one trivial value. Either there is literally nothing of any value whatsoever outside God (although there are some things in some way other than God), or 'God' names all that is.[2] Similarly, there can be nothing truly independent of God (conceived as that which has no limitations); if there were, then there would after all be limits to the Infinite. Spinoza accordingly believed that both 'God' and 'Nature' named the one true, self-sufficient substance. We might speak of It (even Spinoza speaks of it) with terms drawn from our ordinary experience of finite things, but It is really no more like a human person than the Dog-Star is like a domestic dog. True worship means that we must recognize that we are only modes of that one substance, that we must put aside our superstitions and conceits. True worship is the realization that God and Nature is at once the sum of truths, and of all real things: simultaneously everything that is the case (as that it is sunny in north-west England on the evening of 17 July 1996) and everything that exists (such as the black and yellow plastic frog upon my monitor).[3] But neither Spinoza nor the Stoics before him concluded that we should respect 'the dignity and sanctity' of other living things.

On the contrary, it was Stoic thought that persuaded Augustine and other Christian thinkers that non-human creatures were available for any natural, human use. Spinoza (and the Stoics) could, of course, agree that it would be against our real interest to devastate the world we live in. They might even agree, occasionally, that it was in our real interest to appreciate the subtle beauties of the living things with whom we share, and constitute, that world (see Sessions 1977; Lloyd 1980; Naess 1990). But both Spinoza and the Stoics were convinced that it was our duty to be *human*, and that sympathy with anything 'sub-human' was a moral fault. That we should worship the Totality of things, and 'live in harmony with Nature', did not mean that we should respect the interests of other non-human individuals, or take lessons on how to live from what those others did. Nor did they urge us to 'respect' the beauty and integrity of the living Earth (as though anything we did could alter that).

So Spinozistic pantheism (which is what 'pantheism' meant for our eighteenth-century predecessors) is not 'environmentally friendly', any more than Stoicism was. It is true that neither Spinoza nor the Stoics would approve of many uses to which we put our neighbours and the neighbourhood: both believed that many of our impulses were not those of a truly rational being, of one who understood her place in Nature. But they were equally – or identically – convinced that 'being rational' required us not to think that the immediate goals of animal activity were worth respecting. Food, sex and shelter might be what we wanted, but the wise pursued such goals only because, and as far as, they believed that Nature 'wished' them to. What Nature 'wished' above all, for us, was that we live as rational beings, respecting all and only rational beings, and putting aside all superstitions and taboos that might prevent us 'being rational'. So, far from offering a solution to the crisis of our civilized society, such pantheism is one (though only one) of its causes. It is true that Spinoza gave no support to the occasional Stoic notion that 'the world' was made for us (or for the wise among us): Spinoza's universe does without final causes. But there is little real difference between thinking that we are entitled to use things just as if they were for us, that there is no moral obstacle to such a use, and thinking that they really were 'for us' (see Clark 1995).

Spinoza is not the only 'pantheist' or 'sort of pantheist' to be misjudged. There are also pantheists that believe in progress. One Christian thinker regularly invoked as (almost) a deep ecologist, Teilhard de Chardin, believed that humankind (and maybe even one racial subdivision of humankind – and not Australian aboriginals or Bushmen) was now the central line of natural history, that everything in nature should be taken up and humanized, and even Westernized. So, far from supposing that we should respect the given natures of all other creatures, or their sanctity, his thought is all of absorbing or disposing of them. 'Everything precious, active and progressive originally contained in that cosmic fragment from which our world emerged, is now concentrated in a crowning noosphere' (Teilhard de Chardin 1965: 203), which is to say, in humankind, and chiefly in the West (ibid: 233–4). Everything that matters is to be brought together at the End of History, but it will matter only as material for that human or superhuman synthesis. Respecting 'nature' or non-human products of past evolutionary process is, at best, to acknowledge past successes, which must be surpassed if 'Nature' is to be transformed. Whereas Stoic or Spinozistic pantheists expected no strange alteration in the world, no moment when everything thereafter would be wholly different, Teilhard de Chardin (and other even less Christian thinkers) have supposed that there is a real direction, and progression, in the world. Whether his God is, identically, the world, or only *in* the world, it does not follow that the present world or its inhabitants deserve respect. On the contrary, such progressivists regard the old idea that we should respect the gods' own handiwork as superstition. Who are we, after all, but God, or God-in-embryo?

Pantheists may revere the present or the future Unity of all things: the former, more familiar sort could be called Eternal Pantheists; the latter, modernist variety are Progressive Pantheists. Either way it is all too easy to infer that those who recognize the Unity are those in whom that Unity is chiefly reflected. Because human beings can learn to appreciate the world as a single, divine whole of greater worth than any individual element within it, they themselves are judged of greater worth than any other individuals. Those who would appreciate the Unity will usually equate the (human) mind that can appreciate it with the mind of God itself. Those who realize that they, individually, are doing exactly what Nature makes them do thereafter do everything they do exactly for that reason: the mind of the wise is also the mind of God. Having the same mind, or being of one mind, the wise can recognize themselves as the real point of everything: the real owners and rulers, for whom the world, and every thing, exists.

All this is to say that the charges brought against theism by environmentalists of Toynbee's mind can also be brought against pantheism. Theists believe that there is something more worth worshipping than the world, and pantheists that there isn't. But it does not follow that pantheists give individual elements of that one world more respect than theists do. Theists may believe (but need not) that the world here–now is only at best a training ground, whose importance lies in what it does for people. Pantheists may equally believe that things here–now only acquire their value by their present or future integration in a human whole, their hominization or civilization. Or else they may believe that human beings have as great a liberty as any other particle to do exactly what comes naturally. Past and present theists have often been 'environmentally unfriendly'; so have past and present pantheists. Nothing in Plumptre's history of Indian, Greek or late European 'pantheism' (Plumptre 1878) suggests that any pantheists felt any need to care for 'the environment', or for the creatures that we share it with. Nothing in the past or present of Japan, or of any other 'Eastern' nation, influenced by Shinto, Taoism, Buddhism or Vedantin Hinduism suggests that it is only 'theists' who have 'licensed human greed'. Maybe 'the West' has been more successfully rapacious for the last few centuries, but not because we have been careful Christians! To suppose otherwise is actually very strange. The last two centuries have seen the expansion of the 'Western science and civilization' that radical environmentalists most often blame for pollution, exploitation of our neighbours and the deliberate destruction of habitats. But it is 'pantheism' that has been the dominant 'Western scientific' creed: the belief that the 'natural world' is self-explanatory and complete, that people are only and entirely products of natural evolution, and that there is nothing 'beyond' or 'over against' that world by which it – and our behaviour – is found wanting. Not all such naturalists have regarded themselves as 'pantheists', since not all have thought that 'worship' was appropriate at all. But any other religious form can bind us, so the dominant opinion has been, only to fictions: the *real* world is the natural world, and right reason and religion alike must bind us to that nature. This has not stopped our depredations, nor is there any clear reason why it should.

There are two further problems for the pantheist, even if she avoids the peril of equating the Divine with her own knowledge of it, and so assigning greater value to humanity on the paradoxical plea that humans can appreciate that there are better things than us. First, it is all too easily assumed that because God and the world are one, it must be obvious that this world here, this Earth, is God. But the (maybe infinite) Totality[4] is not the same as Earth. The Earth, even the living Earth or Gaia, is only one small segment of the Whole. What is to say that Gaia (that is, the coordinated sum of living things on Earth) is more significant, more typical, more central to the Whole than the dust of a distant nebula? Even if pantheists' worship of the Unity required them to care for it, or not to damage it (and how exactly does one damage the Totality, or help it on its way?), why should it follow that we should not damage Gaia? And how, in any case, could we? Everything we do (or seem to do) is what God and Nature requires: even if, by human action, the terrestrial biosphere is seriously simplified, or even ended, no damage will have been done to the Unity, nor even to the biosphere (any more than we as individuals are damaged by maturity, or change, or death). If it is the Totality that is divine, our worship of it does not mean that we should or could sustain it against its enemies (for it has none). If Gaia itself deserves respect as being a local manifestation of the Totality, it does not follow that we should hope to make it live for longer than in fact it does, nor that we should keep it in any one particular condition. From our original, human point of view it may seem that the spread of deserts damages the Earth: from a less personal point of view, this only changes the environment, and opens it to other forms of life that are as much a part of the Totality as us.

All this is obvious enough. Pantheists (or at least Spinozists) may agree with atheists that the world was not created for our special benefit. They may insist that we are no closer to, and no more like, the divine than weevils, mitochondria or quarks. If they hold firmly to that view, it is not clear that any particular state of Gaia is more to be desired or esteemed than any other: pollution, environmental degradation, species extinction are of no concern to Gaia, or the Divine Unity. If, on the other hand, pantheists follow precedent, and claim greater value (as the mind of God) than all the other elements of this world here, their position is, exactly, that of civilized humanity – and they give no greater dignity to other creatures than, so Toynbee claimed, the Judaizing monotheists. Stoic pantheism is, if anything, historically to blame for humanistic arrogance, for the claim that nothing matters but the human mind, and what can do it good. Either things as they are are perfect (and there is no crisis), or things as they shall be are perfect (and the crisis is a step upon the way, part of the hominizing, civilizing process): we may be 'sorry' for the losses on the way, but should not let ourselves be guided by irrelevant emotion.

ROMANTIC PANTHEISM

But this cannot be all the truth. Those who call themselves pantheists (and those who hope for a pantheistic revival) may mean something different. Philosophical Pantheism (whether Eternal or Progressive) may not be the 'pantheism' that pagans and environmentalists intend. One first step away from Spinozism is *romantic* pantheism: 'the universe is a living unity which could be known through the imagination' (Piper 1962: 3). Those who worship *that* Unity must give more weight to 'the imagination' than to 'reason'. Reason, on this account, can know only the bare bones of things. Reason, at least if we equate it with our current scientific theorizing, must have doubts about the real *unity*, present or yet to come, of a cosmos which has been out of touch with most of itself since a few seconds after the Big Bang.[5] Romantic imagination, fuelled by love, reveals the inwardness of the one power that moves through all things. Those who think of rocks, or flowers, or fish, or sheep, or people only as they seem 'from the outside', as particular entities, moving in accordance with mathematically discoverable laws, see only half (or less than half) the story. Seeing them with the eyes of the imagination, through romantic sympathy, is seeing their significance, their shape as wished-for forms of the One Spirit. Such romantics are a little less likely to end as anthropocentrists (though they may be charged with undue anthropomorphism). If the Unity is revealed through reason, as an interlocking system of cause and effect, then (simultaneously) everything that happens must be known to be inevitable, and the *human* mind is where the Unity is known. If the Unity is revealed instead through empathy, and so through the experience of our own freedom, then (simultaneously) there is nothing that *must* happen, and it is not the *human* mind alone that can experience it.

If philosophical pantheism is likely to end in one form or another of deterministic humanism, romantic pantheism is likelier to end as 'nature mysticism'. Love of 'Nature', and especially of the nature that is manifest within the Living Earth and all its denizens, will usually distinguish the natural and the human. Even though human beings, too, are 'natural', it is (uniquely?) possible for human beings to forget the fact. By reasoning our way to power, by laying out Cartesian grids in streets and national boundaries, we hide from ourselves the true continuum. Out in what we can conceive as 'wilderness' (or even 'countryside') we can forget what makes us 'different', 'detached' or 'alienated' from the Whole. Cities, in one sense, are as natural as ant-hills, and as likely to contain a host of unseen agencies. There is nowhere (and will never be) a region sustained 'against nature' by the will of human architects alone. But in the city it is easy to forget the natural powers sustaining everything we seem to build. In wilderness (even one infected or constructed by our human actions) we find it easier to recall the truth. Nothing that we can ever do (not even poisoning the seas and cutting down rain forests) can ever alter the reality: the Spirit of the Unity will still be present everywhere, and still be visible through the Imagination. But those who know that Unity will never wish to poison seas or cut down forests: only the deluded can imagine that they (as individuals) will survive that chance; only the deluded can imagine that a landscape 'ordered' from outside is more beautiful than one developed from within.

This last contrast is one that would-be pantheists will emphasize. Creation myths will frequently describe a god's imposition of order on disordered stuff. Ruether (1993: 19), among others, has suggested that even the Hebraic story (which is much less mythological than most) has merely replaced the Babylonian monster, Tiamat, slaughtered and dismembered by Marduk to construct the world, with God's immediate ordering of a malleable stuff. Plato's influential story (in the *Timaeus*) of the creator's fixing of form in matter, she says (ibid.: 23), also demotes the world to the inferior status of an ownable, made thing. By implication, if we thought of the world as ordered from within we would recognize it as something not to be controlled or owned or ordered 'from outside'. Believing it to be constructed, ordered, from outside (and in the Platonic version, by a rational intelligence principally ascribed to males), we will think it is available for our possession and our use. We can join with Mind in making something orderly from the mere rubble, and should. Accordingly, so pantheists claim, it is best – if we are to treat the world decently – to conceive it as a living creature, not a construct. Constructs (and slaves) live by another's life, and have no real being apart from that primeval other: living creatures, and the world, cannot be tools or slaves of any other creature. Either they 'own' themselves or they are owned by the same power that moves in us. Making a slave or tool of anything at all is to deny ourselves.

The claim that *Plato* is to blame for our alienation from the living world, and for our tendency to tyrannize, is not one worth sustaining. Historically, the same poets and philosophers who have been claimed as romantic pantheists, or deep ecologists before their time, have usually, in fact, been Platonists (see Armstrong 1976; Clark 1997). Plato treats it as an axiom that decent rulers are devoted to the welfare of the ruled, and also that it is God (not us) that is the real ruler. But Ruether's idea may still prove fruitful. Theism is the thesis that there is something well worth worshipping which is distinct from every individual finite thing, and even from the Totality of finite things. Pantheism, on those terms, is the thesis that there are no finite things distinct from what is worshipful. There is nothing, that is, existing before form, before the Spirit moves in it: and therefore there is no question of 'imposing' form on something not itself. The *Timaeus* story, as all readers knew, is only a way of speaking, and must not be taken to imply that anything is a 'made' thing in the sense that Ruether means. Conversely, nothing at all exists without the Spirit, and everything is thus dependent on that Life. But, of course, it is now unclear how theism and pantheism differ. There is, for pantheists, a real distinction between individuals considered singly and the Spirit that sustains and links them. There is, for theists, an immediate recognition that nothing at all exists independently of God. Both must agree that nothing that we see can ever be anything but God's action in that particular case and situation. Both must agree that it is possible, paradoxically, for finite creatures to forget that truth, and so require the Spirit to move in ways that otherwise it wouldn't. The only problem that remains is, simply, how best to recall the Spirit or the Life that moves in all things, and the Form that all things move toward. Shall we conceive the Thing we worship as Another, by whom we may be moved, or as the only Unity, to which we may aspire by putting on one side the sense of individual, rival being? If as Another, that Other may display itself to us in the face of our fellow creatures. If as the Unity, we may be reminded of It by our sympathy with those same fellow creatures. Either way, we are reminded that our ordinary selves are less than the self, the society, that may exist.

Is this to say that theists must, in a way, be pantheists? Theism requires, indeed, that there are in the end no other powers than God, nor any independent truths. The argument I sketched before, that God must include everything of value, and be the only self-dependent real, is almost orthodox. 'Everything that happens is a sacrament of God's will' (Cardenal 1974: 105). Even the claim that pantheists dispense with any 'merely personal divinity' is not much to the point: the theist's God is not a being among many, nor a personal being like us. God's love and anger are not what love and anger are in us, even if our 'love' and 'anger' get their names from His. A pure monotheism might be difficult to distinguish from philosophical pantheism: was Jonathan Edwards (America's first original philosopher, and a hell-fire preacher) Calvinist or Spinozist? If he avoided pantheism it was, paradoxically, because Christian theism identifies *one* human being as God: God, or at least God's Word or Outward Face, identically *is* a human individual. In that sense, God is, after all, 'a being among many', 'the firstborn of many brethren' – but by the same account, 'the incarnation of the Word in a human body means his incarnation into the whole cosmos' (Cardenal 1974: 137). The whole cosmos is 'God's Body' not because the cosmos secretes God (as the human brain, on modernist accounts, secretes thought) but because the Spirit that allows all things their life has 'chosen' to experience that life as they do, and so enabled them to 'experience' Himself as He does. He has become Man (as Athanasius said) so that we can become God. Now or someday the totality is brought together into a single whole: not only united in its single origin, but in its synthesis – a synthesis not dependent on merely physical connections, nor such as to cancel individual being. God, so Tibetan lamas told Francisco Orazio, is the company of all the holy ones (Kant 1970: 107).

POLYTHEISTIC PANTHEISM

But that may be the real point at issue. Theists and philosophical pantheists believe that there really is a universe, a unitary whole. Theists emphasize that the life of that whole is something distinct from it, something that 'could' exist and still be no less perfect even if no finite thing had ever been 'created'. The real being of every finite thing is There in God, and for that reason there is a demand imposed upon creation, a law that is not just equivalent to whatever happens here. Pantheists of the familiar sort emphasize that nothing can exist outside that divine Unit. Every finite thing is just a mode, or version, or refraction of one excellence; everything that happens does so 'by God's will.'[6] Theists are often tempted to believe that they know what Good it is that the totality of being expresses (now or in some future). Pantheists are more likely to believe that the totality, God, has no distant purpose, and does not intend 'our' good. But both agree that there is or will be a perfected form, a unitary whole, a single something that deserves our love. Either may perversely conclude that it is our task here to hurry on that consummation, that we should always act to bring about that good (or that whatever accidentally we do will, by the will of God, advance that cause, and so we needn't worry).

But how can there be any such Unity *of consciousness* in the world as currently conceived? Even a Unity of material connection is debatable:

there is reason to think, as I remarked before, that there are regions of the natural world that have been wholly inaccessible to each other since the Beginning (and will always be). And if *my* conscious mind is any sort of real thing, it is not *yours*: the God that is both of us apparently exists without any Unity of consciousness, or else (if It shares the ownership of the contents of our otherwise distinct mentalities) it is insane (see McFarland 1969: 175). Theists insist instead that God, although the true life of all that lives, is still distinct from every creature here. 'The reality we see is as unreal compared to the reality in God as a coloured photograph compared with what it represents' (Cardenal 1974: 83). Another sort of pantheist abandons the claim that God is One at all.

Some pantheists, remember, think that the Whole is God: these are the 'philosophical pantheists', and even the 'romantic pantheists', whose doctrines I have doubted. Other pantheists mean rather that every single thing is divine, not that there is one thing, the greatest of all things, that is God. Such pantheists, less confusingly, might actually be called polytheists, or pluralists. They may not even think that there is *a universe* at all: there may not be a single, unified totality, now or in the future.'[7] Even if there is, it is not *that* which is the only God. Rather 'every thing that lives is holy,'[8] and is not devalued because it is damaged. Whatever we encounter we should love and honour, and also (obviously) recall our own love-worthiness. It need not follow that it is *the same* god we encounter everywhere: there are as many gods as there are real things, and all of them divine. Such gods are not, severally or corporately, 'that than which none greater can be conceived'. The point is rather that no better maple-tree-outside-my-window can exist, no greater cat-with-almond-eyes-beside-my-desk or green-and-yellow-plastic-frog. Each real thing deserves our worship as itself. Nothing can be compared, and found wanting, with any other thing: the god that is an individual human child is not a greater god than one that is a tree, a rock, a microbe, or a cell within that child. 'God is in all creatures, man and beast, fish and fowl, and every green thing.'[9]

Does it follow that we should not distinguish between things at all? Does it follow that we must think all things are perfect as they are, whatever that may be? Sunrise and shipwreck, cowslip and cancer, live dog and dead lion, all equally divine? If that were so, the respect environmentalists and good 'pagans' require us all to give to 'healthy' habitats, clean streams and active wildlife should not be any greater than we give 'unhealthy' or decaying systems. Like philosophical pantheists we should admire things 'as they are', and reckon plague and famine as no more than charming changes. But 'pagan' pantheists may answer that it is one thing to love, admire and worship the divine, and another to believe it perfect. The divinities such pagans find need not be thought invulnerable: on the contrary, the gods display their godhead in being changeful, weak or ill. Helping a god may be what gods do best. Here, too, it is not quite clear how 'pagans' and theists differ. Theists may also say that every real thing is good: 'evil' lies only in relationships, or in a clouded vision. Even the smallpox virus (which is regularly mentioned as an entity, or class of entities, that 'ought' to be extinguished) is no more 'evil' than the creatures who attack it. An unclouded vision would sometimes see that sickness or ill health do no real harm; an unclouded power would sometimes find a happier, more symbiotic way for living things to live. Both theists and 'pagans' will, contrariwise, agree that it is often good (or godly) to seek that symbiosis, or, failing that, to heal or comfort where we justly can.

There is one point especially that theists and 'pagans' can agree upon, against the prevailing idol of this age. Secular moralists will often speak of 'lives that aren't worth living', 'worthless lives', and urge us not to create or prolong or even allow such lives. It is a rational conclusion: a life stripped of all goods (of health, wit, pleasure) is not much worth respecting, and if there is no self distinct from its experience, there is no good thing there to honour. Only if a self can have or be a value in itself, apart from its experience and qualities, does it make much sense to think of it as worth respecting when it has no worthy properties. Why, superstition apart, should we or could we 'respect' foul, stupid, miserable tramps? Making those lives less miserable is likely to be costlier, in many ways, than just eliminating them, and making less miserable lives. Only if there is something there that demands or requires or 'merits' love, in the absence of any good apart from its own being, does it make much sense to give it love. Aristotle, when attacked for giving charity to an 'unworthy' object, replied that he had not given to 'the man' but to 'humanity' or to 'the human thing'. We have learned, and still in part remember, to respect and love the human form as being 'divine.'[10] All 'pagans' ask of us is that we see all forms alike, and wish for them that they can visibly be the gods they are.

The polytheism that identifies 'gods everywhere' differs from a 'merely moral' respect for every individual in that we can draw power from those gods at the same time that we offer them our help. The very same individual that, in one way, needs our help – or at any rate our forebearance – must also be conceived as helping or inspiring us. It is not an unfamiliar paradox. Consider the role of children: it is they, as often as (perhaps more often than) our parents, who excite 'religious' awe, even though they are most obviously vulnerable. Children are sacred precisely because we can hurt them. And though we can hurt them they – or the 'innocence' they represent – are undefeated.[11] 'Pagans' ask us to look with something like the same sensibility at every creature we encounter. That gospel is compatible (how could it not be?) with our feeling special loyalty to our own local gods (including human children).

Pagan polytheism, therefore, is still compatible with much the same discriminations that we ordinarily make: where it differs from its theistic or atheistic rivals is in its denial of 'objective' order. Our gods may matter more to us than other gods, but only because they're ours: others may and will have other loyalties. Whereas an ordinary humanist is likely to believe that everyone, or every rational one, must honour *people* more than rats or fishes, a decent pagan is more likely to behave like a decent nationalist. *My* nation matters more to me than yours, but for that very reason I expect your nation to matter more to you than mine ('My nation matters more to me' is a claim affirmable by anyone at all). Does this constitute a problem for environmentalists? We can imagine peace in heaven: maybe some day the gods agree at least to differ. But our actual experience is war. Not only do the gods compete for space and loyalty, it must sometimes seem that their very godliness depends on competition. A lion that lies down with lambs is not so very lion-like. Being a *divine* gazelle (which by hypothesis each gazelle is) may still demand being eaten by an equally divine lion. Perhaps all that can be said is that the 'war', to be compatible with pagan sensibility, must still allow each warring entity 'to be itself'. There are gazelles because there are also lions: being a gazelle requires the risk, the opportunity, of being eaten. Those who eat and are eaten pay the price of being.

One further twist to 'pagan pantheism' (or polytheism) is given by the thought that, though an individual may die, its being the thing it is endures, as childhood innocence outlasts the death of children. Individual fishes, each of them a god, may seem to perish, but 'they' or their equivalents or even just 'the fish' return. Presumably that feeling lies behind the widespread intuition that it is worse if a whole 'species' perishes, since it is not clear how then the dead return. The thought is there – but it must be weighed against the other thought, that no real entity is just a sample of some type. Whatever individual is lost is gone, even if there are others who can 'take its place'. That loss – unless we revert to Eternal Pantheism (for which nothing ever can be lost) or orthodox Theism (for the God remembers) – is absolute (and part, again, of the peculiar divinity of finite things).

I conclude that the 'pantheism' Toynbee and so many others have desired is not the philosophical version, but the 'pagan'. The point is not that there is or will be a Totality, the one real substance greatest to be conceived. It is not even that the Whole is understood through imaginative sympathy, not careful reason, though pagans do themselves refer, at times and aphoristically, to 'the Divine Oneness, the Unity of All'.[12] The point that such revivalists require is simply that 'everything that lives is holy': whatever we encounter is 'a god', precisely because it is vulnerable, finite, mortal. On the one hand, losses here are always compensated by the coming into being of other, fresh, new gods (or the return of old). On the other hand, every loss is real, and reveals divinity in our remembrance of it. On the one hand, any casual or ungrateful use of these our fellow gods is only possible for those who have forgotten godhead: 'We offered our respects and gratitude to the fish and the Sea Gods daily, and ate them with real love, admiring their extraordinarily beautiful, perfect little bodies' (Snyder 1969: 139).[13] On the other hand, we are bound to use them in the ways required of their being (and must expect ourselves to be of use). Gods should not be used ungratefully, but it is the duty of gods to give themselves for use. For polytheists or pluralists there is no single World, and never will be one: there is no unity beyond the mere continuum of warring partners. Unfortunately, even this vision is inadequate: those gods that are the expressions of our very human imagination may, as Blake realized, defy 'the infinite and eternal of the human Form',[14] by losing any allegiance to a unifying, sacred power. But they cannot defy the human heart: they *are* that heart, projectively. It follows that the demands they make can never be for real reform. The gods are idols, even if (for some) they reinforce their hearts' desire for peace – a peace impossible as long as there is war in heaven, and in the human heart.

CONCLUSIONS

Pantheism has been the dominant ideology of Western civilization in the last few centuries. On the one hand, there is (or we have thought there was) a single, natural universe. That universe is closed to outside intervention, and all the principles needed to explain it are to be found inside it. Those principles turn out (or we have thought they turned out) to be mathematical: whatever happens in that single, real world is the result of mathematical transformations of theorems of the expected 'Unified Field Theory'.[15] Human beings are simply products, rather recent products, of that single system, and have no higher duty than to do whatever 'comes naturally'. On the other hand, that single, centreless universe is not one in which we, as human beings, can seriously live. It is inevitable that we will continue to see things in relation to our selves and histories. Even though we 'believe' in an impersonal, uncentred universe, we shall live in one animated by memories and omens, ghosts of the past and future. This is the realm of polytheistic 'pantheists', who need not (and in this age of the world almost certainly do not) believe that the gods are 'objectively' real. It is enough that they fill our subjective spaces.

Philosophical pantheism (whose chief form is scientific naturalism) is a denial of anthropocentrism. Objectively, we are not the most important creatures in the world; we are not the goal to which creation moves. As Spinoza demonstrated, this need not imply that we are morally or logically bound to give other creatures any consideration. Poetic or polytheistic pantheism allows us, without realistic belief, to acknowledge the subjective force of gods, ideas, symbols, and affections. Some such pantheists undoubtedly feel more deeply for our fellow creatures because they can see them as gods, or see the gods in them. But in the last resort such godly forms are known to be projections, and offer no real barrier to our use of 'the environment' and the creatures that we share it with. The gods that we see are those that validate the use we wish to make of them. Objective and subjective pantheism alike can constitute no real challenge to our 'natural' impulse, which is to make as much use of the world around us as we can. Those who made 'pantheism' a term of theological abuse were, after all, correct: neither the objective universe (as currently conceived) nor the subjective realm of nature-gods and animating spirits is an adequate response to the mystery of being, and our duties as finite creatures in a world we did not make. *Pace* Toynbee, what we need is the one true rival to all forms of pantheism (that is, the very creed he said we didn't need: see Clark 1993).

NOTES

1 See Levine (1994: 17n2); McFarland (1969: 266–7).
2 Strictly, if God is infinitely valuable, no finite addition can increase that value: $\infty + n = \infty$. One of the things that thinkers of Spinoza's day found difficult was that any proper subset of an infinite set was also infinite: $\infty / n = \infty$.
3 The usual way of describing this conjunction is that God can be conceived both as mind and as matter: *truths* only exist as mental objects ('true propositions'), and there can only be a real sum of them by being integrated in (necessarily) an infinite intellect. But this Mind is nothing like our own fluctuating, incomplete mentality.
4 Whether that natural Totality is infinite is moot: there are good scientific reasons to suspect that it isn't, notably that, if it were, the light and gravitational attraction at any point (say, here) should also be infinite (see Jaki 1993: 6).
5 Most of the cosmos has been since then, and always will be, out of our reach: nothing going on Over There can influence events Here, since Here and There are and always will be too far away for light from one to reach the other. It is not even certain that this world here, this Earth, is any real Unity.
6 The distinction between 'substances' and 'processes' which some post-modernists and pantheists deny is actually vital: only if there is a distinction can we coherently say that every thing that lives is holy, but not everything it *does*.
7 In the New Age, so Yeats glumly or gladly prophesied, we would worship many gods, not one alone, and 'receive from Joachim de Flora's Holy Spirit a multitudinous influx' (Yeats 1962: 392; see Clark 1989: 30*ff*).
8 William Blake 'Visions of the Daughters of Albion', plate 8, line 10 (1793) (Blake 1966: 195).
9 Jacob Bauthumley (*c*. 1645), cited by Thomas (1984: 301).
10 'And all must love the human form, in heathen, turk or jew; where Mercy, Love, & Pity dwell there God is dwelling too', from 'The Divine Image' (Blake 1966: 117).
11 Which is one reason why we find it horrifying to consider their corruption: both the abuse of children and the occasional cruelty of children appal us.
12 See, for example, Selena Fox's credo 'I am a Pagan' (http://www.iag.net/~dakins/pagan.html).
13 Though, as Aristotle said, those who 'love' wine do not wish it well, but that they should drink it. It is not altogether clear that we can be truly 'grateful' when we never bother to repay the debt (and how could we repay the fish?).

14 William Blake, 'The Four Zoas', night 9: line 367.
15 There is at present no real way of deducing mere existence, or even the brute facts of how things went in the first beginning, from a mathematical equation. Theists may conclude that finite existents, like the Universe, exist because God makes them. Philosophical pantheists prefer to believe that they will one day be able to deduce everything from necessary theorems (and raise no question about the being of those theorems).

5 The real, the one and the many in ecological thought[1]

Freya Mathews

PART I

One evening a couple of years ago I was driving home through city traffic into the sunset. With all the objects around me so finely and blackly etched against the orange light, the differences between trees and tele-graph poles, birds and distant aeroplanes, no longer registered. I was filled with a sense of one of those semi-ineffables: that every object, every instance of matter, is not merely manifest and visible, but actually *there*, present to itself. It 'feels' itself, not in the sense that we feel heat or sharpness or pain, but in the sense that there is an innerness to its reality as well as an outerness. Such a sense of the innerness of matter normally eludes us. Its elusiveness is what gave rise to the appearance/reality distinction in traditional philosophical thought: how to distinguish the 'appearance' of an object from the 'reality' of it? How to make any sense of the claim that the world as I perceive it is real as opposed to a mere dream or hallucination? The traditional answer to this question was that the world is real to the extent that it is concrete or substantial, and it is concrete or substantial if its properties are grounded in, or 'inhere' in, some kind of substrate. However, if we try to give content to this notion of substrate, we find that we cannot do so – it is simply understood, in circular fashion, as that which makes an object real as opposed to merely phenomenal. As Berkeley showed, no empirical account can be given of the real world that would distinguish it from an order of mere appearance.

The difficulty of providing an account of the reality or concreteness of the world is echoed in the classical problem of solidity. Solidity was traditionally seen as the – primary – property that distinguished a material body from an unoccupied stretch of space with the same appearance, that is, the same shape, size and, perhaps, colour. It was the solidity of the body that assured us that it was not a mere phantasm or illusion. However, unlike other primary qualities, such as size and shape, solidity could not be characterized in intrinsic terms: there was nothing in the body itself that was in any way describable by us that rendered that body solid. Its solidity could only be defined extrinsically or relationally, in terms of impenetrability – as the capacity of the body to keep other bodies out.

But as an account of the concreteness of a body, of its actual occupation of space as opposed to its mere appearance of doing so, this is clearly question-begging: the body in question will only qualify as solid if the bodies it keeps out are themselves already solid. There is, in principle, no reason why an order of illusory bodies should not be such that those bodies appear to keep each other out; their doing so, however, will not render them solid (think of the cinema). In other words, the definition of solidity in terms of impenetrability only works if the body to which impenetrability is ascribed is assumed to belong to an order of already solid bodies.

Neither classical nor post-classical physics has solved either the problem of solidity in particular or the appearance/reality problem in general. Solidity itself, of course, never appeared as a variable in physics, but analogous properties which did, such as mass and charge, are disposition-ally defined, and as such they too may be question-begging as accounts of what it is for a body to be really *there* – what it is for a body to be real as opposed to apparent or illusory. For when the inertial mass, for instance, of a particle, p, is defined as the disposition of p to resist acceleration when a force is applied to it, the notion of disposition is in this context generally nothing but a reification of the empirical conditional that particles with the empirical properties of p will in fact resist acceler-ation to a particular degree when forces are applied to them. In other words, the notion of disposition is in this context vacuous: it in no way lifts the notion of mass out of the realm of appearances – it provides no idea of a concrete something which causes, produces or underpins the behaviour in question.[2]

So it seems that – as philosophy has tirelessly attested – our senses can never reveal to us that which gives a body its concreteness. We can see its colour, feel its impenetrability, and so on, but there is a sense in which these are surface qualities only – mere appearances. The inner reality of body – the 'thing-in-itself' – is never revealed. The acknowledgement of this has left a sceptical legacy in philosophy: the world is conceived as façade only, as a parade of appearances. Belief in its evident reality is suspended. Matter becomes ideal, or phenomenal, or socially constructed (depending on the philosophical era in which you happen to find yourself). The idea that it may be present-to-itself, independently of whether it appears to us, barely arises, for it is assumed that such presence-to-itself is unrepresentable, and hence beyond the horizons of our imagination.

But that night a couple of years ago, driving into the sunset in the peak-hour traffic, I had a sense of the world from within, a sense that everything that exists in the realm of extension – telegraph poles, over-head wires, factories, roads, billboards, tyres – is present-to-itself. What might such self-presence consist of? Perhaps it could best be characterized via an extension of, or analogy with, the notion of subjectivity. But then what is subjectivity? Subjectivity in us is, of course, associated with consciousness, and in other beings with sentience. However, it is not identical with thoughts, feelings or sensations, but rather subtends them. Subjectivity is that deeper level or field of self-presence out of which thoughts and feelings arise. We are – *contra* Descartes – alive to our own corporeality even when we are not thinking at all: our flesh is present to itself whether we are conscious or unconscious, awake or asleep.[3] That is to say, our bodies go on existing for themselves even when they are not existing for us – when they are not being registered by our conscious minds. It is perhaps by analogy with this unconscious subjectivity of living flesh that we might understand the subjectivity of matter in general. For by imagining the way that our sleeping bodies are present-to-themselves, we can perhaps imagine the way that matter generally is present-to-itself: just as the sleeping body is not a purely externalized object, but occupies space from within as well as from the point of view of an observer, so all bodies may be imagined as occupying space from within in this way. By saying that objects occupy space from within, I do not mean merely that objects have internal parts, for these parts are themselves normally imagined or conceived as externalized, or as they would appear were they exposed to the view of an external observer. For an object to occupy space from within in the present sense is for it to possess a dimension which is, like subjectivity in our own case, in principle invisible or otherwise insensible. And just as it is our subjectivity – the innerness or presence-to-itself of our own bodies – that assures us that we are really here, that we really do occupy the space that our bodies appear to occupy, so it is this innerness, this presence-to-itself, of matter generally that renders the world at large real as opposed to mere externalized husk or insubstantial phantom. From this point of view, 'subjectivity', in an extended or analogical sense, is the elusive property that distinguishes the thing itself from the mere appearance of it: it is the fact that bodies are present-to-themselves – that they occupy space from within as well as from without – which ensures that they are really *there*.

However, does this idea of unconscious subjectivity really stand up? For one classical line of elucidation, let's turn to Leibniz, who appeals to such an idea in characterizing his 'simple substances' or monads. Monads, which manifest (indirectly, via the pre-established harmony) to other monads as material things, are characterized by Leibniz in *The Mona-dology* in terms of pure activity, where mind provides the model for such activity – thoughts, or, in his term, 'perceptions', flow imperceptibly one from another, without need of external cause or stimulus. That is, although these 'perceptions' are, in the case of any particular monad, of bodies, particularly the body associated with the monad in question, they are not directly caused by any external or extensional order of reality, since for Leibniz no such order exists; rather they are implanted in each monad, and fortuitously synchronized with the 'perceptions' of other monads, by God. Leibniz states, 'there is nothing besides perceptions and their changes to be found in the simple substance. And it is in these alone that all the internal activities of the simple substance can consist' (Leibniz 1973: 254–5). To ascribe 'perceptions' to monads generally is not, however, to imply that they are necessarily capable of the kind of sensory experiences that we enjoy. As Leibniz elaborates: 'We experience in ourselves a state where we remember nothing and where we have no distinct perception, as in periods of fainting, or when we are overcome by a profound, dreamless sleep. In such a state the soul does not sensibly differ at all from a simple monad. . . . Nevertheless it does not follow at all that the simple substance is in such a state without perception. . . . When, however, there are a great number of weak perceptions where nothing stands out distinctively, we are stunned; as when one turns around and around in the same direction, a dizziness comes on, which makes him swoon and makes him able to distinguish nothing' (ibid.). In other words, in the simple substances, perception is so confused as to amount to nothing more than a grey fog, and in this sense simple substances may be said to be pre-conscious and pre-sentient, even though endowed with subjectivity.

Leibniz then posits an unconscious form of subjectivity, and attributes it to simple substances. I am far from wanting to advocate a Leibnizian metaphysic here,[4] nor do I even want to suggest that the 'subjectivity' of matter should be understood as a dull or confused form of perception. The theory of unconscious subjectivity that I shall be developing in the rest of this chapter turns on a notion of conatus rather than perception. But Leibniz helps us to gain some imaginative purchase on the notion of an unconscious form of subjectivity associated not only with animate but with inanimate things.

I should also point out that to explain the reality or concreteness of the world in terms of an extended or analogical notion of subjectivity, as I am doing here, is not to commit to the idealist position that the real is *merely* a locus of subjectivity, and not in fact material at all, as Leibniz held. Nor is it to espouse idealism in the Berkeleian or phenomenalist sense – the kind of idealism that postulates that 'to be is to be perceived'. It is rather to suggest that matter cannot be characterized exclusively in terms of extension, as empiricists have traditionally supposed, but must be attributed with interiority as well, where interiority is conceived as analogous to subjectivity.

However, this is a conditional hypothesis, for I am arguing only that *if the world is real*, then it must have a 'subjective' dimension. In other words, this is a point about the *conceivability* of the real, rather than its knowability: if there is indeed a distinction between appearance and reality – if things can exist mind-independently, as well as merely appearing to perceivers – then the only way of making sense of this distinction, or giving conceptual content to it, is to impute 'subjectivity', in something like the present sense, to matter. To qualify existence as 'mind-independent' is not in itself significant – cannot in itself stand as an adequate characterization of 'reality' – as long as the only notion of existence available to us is one that can be exhausted in terms of mind-dependent appearances. Hence physical realism cannot be explicated simply in terms of mind-independence: some way of making conceptual sense of mind-independence itself is also required. My reason for claiming that the *only* way of making sense of physical realism, or mind-independent existence, is in terms of the interiority or 'subjectivity' of matter is that the limits of conceivability are a function of the limits of our experience, and our experience is exhausted by the empirical (the realm of appearance) on the one hand, and the introspective (the realm of interiority, subjectivity) on the other. Since, as we have already seen, empirical experience can provide no conceptual means for distinguishing appearance from reality, only introspective experience holds the potential for doing so, by suggesting that *if* material things are real, they must be present-to-themselves in a way analogous to that in which we are present-to-ourselves.

This is not to claim that the attribution of a 'subjective' dimension to matter in any way helps us to *know* whether the world is real or merely apparent. Since the purported 'subjectivity' of things would not be empirically accessible to us, we could be no more certain that material objects possessed it than we are currently certain that other persons do. It is rather that unless we appeal to such a 'subjective' dimension, or something analogous to it, we do not have the means even to *conceive* of the distinction between appearance and reality, at least in metaphysical terms: this is not a meaningful distinction.[5]

However, although this hypothesis that the material world is 'subject' as well as object cannot, by its very nature, be empirically verified, when we do approach the world as 'subject', especially as a 'subject' potentially responsive to our overtures, then that world may begin to manifest itself to us in entirely new and surprising ways. In this sense – a sense ruled out by the presuppositions of classical science – the hypothesis may indeed be 'testable', though 'testing' it will, if it is true, no longer be the point, since the point will now be, as I explain in Part III, to *relate* to the world, mutualistically, rather than to know it in the traditional, unilateral way.

PART II

The problem that immediately confronts the argument that I have presented in Part I – the argument that imputes a 'subjective' dimension to matter on the strength of the appearance/reality problem – is that matter is no longer, in contemporary physics, a theoretical primitive. That is to say, physical reality is no longer conceived as coextensive with matter. The appearance/reality question can thus be posed in relation to non-material as well as material aspects of physical reality: how are we conceptually to distinguish between real and merely apparent light, for instance? Can we say of the non-material aspects of physical reality that they too have a 'subjective' dimension? And if we can say this, does it alter our notion of the 'subjectivity' attributable to physical reality?

If the argument from the appearance/reality problem is to succeed, I think that we do have to extend it to physical reality generally, rather than restricting it to matter. But this step forces us to consider an issue that was not addressed in the previous section. This is the question of the relation of subjectivity to the *subject*. Subjective experience, whether conscious or unconscious, is, after all, the province of a subject: there is presumably no such thing as free-floating subjectivity. However, a subject, understood in this way as a centre of subjectivity, is necessarily an indivisible unity: there are no scattered subjects, and I think I can say, without being too controversial, that the boundaries between subjects are not nominal (i.e. it is not a matter merely of choice or convention whether a particular set of experiences is ascribed to you or to me; those experiences are already either yours or mine). The individuation of subjects, or centres of subjectivity, is thus an objective matter, an ontological given. Since matter is not on the face of it externally objectively individuated in this way, however – which is to say, since material things are not themselves indivisible unities – we have to ask how the material realm is to be externally divided to correspond with its purported internal differentiation into subjects, or centres of subjectivity.

This question has not been squarely faced so far because we intuitively think of matter as parcelled up into spatiotemporally bounded particles or objects. When we do think of matter in this way, and then ascribe 'subjectivity' to it, it is natural to picture each object as also a subject, or centre of subjectivity, to which the materiality of the object is subjectively present. However, although it may be intuitively obvious to think of matter in this way, it is mistaken: matter is not really parcelled up in convenient packages, and many of our individuations in this connection have purely nominal status. Thus the question of how to divide reality into subjects, or centres of subjectivity, to which materiality is subjectively present, has not yet been dealt with, and this question becomes acute when the argument is extended to the physical realm generally. For there is not even any intuitive presumption that the non-material dimensions of the physical realm – field or wave-like processes, for instance – can be carved up into individual units. Yet if these non-material dimensions cannot be so carved up, how, again, can physical reality be externally differentiated consistently with its purported interior differentiation into manifold subjects, or centres of subjectivity; in other words, how can 'subjectivity' be imputed to physical reality?

One effective way of reconciling the ontological unity and indivisibility of subjects, or centres of subjectivity, with the generally nominal unity of physical objects is to adopt a *holistic* approach to physical reality. If physical reality as a whole, under both its material and non-material aspects, is seen as constituting a genuine, indivisible unity, then it could itself justifiably be regarded as the subject, or centre of subjectivity, to which the entire differentiated physical manifold is subjectively present.

I have argued elsewhere that physical reality as a whole can indeed be regarded as an indivisible unity (Mathews 1991: chapters 2, 3). I lack the space to recapitulate those arguments here. Suffice it to say that if we adopt a substantival view of space, and a geometrodynamic view of physical process, then the universe may be conceived as a unified, though internally differentiated and dynamic, expanding plenum. Since such a plenum is necessarily self-actualizing, and since it is holistically rather than aggregatively structured, it also, according to my arguments, qualifies as a 'self', or self-realizing system. A self-realizing system is, in this context, defined in systems-theoretic terms as a system with a very special kind of goal, namely its own maintenance and self-perpetuation. On the strength of its dedication to this goal, such a system may be attributed with a drive or impulse describable as its *conatus*, where 'conatus' is here understood in Spinoza's sense as that 'endeavour, wherewith everything endeavours to persist in its own being' (Mathews 1991: 109).

As an indivisible unity, the plenum can serve as a subject, or centre of subjectivity, to which the materiality or physicality of the universe as a whole is subjectively present. In other words, the difference between the world as real and the world as mere appearance can be explicated via an attribution of subject status to the world as a whole. Such an attribution is not only rendered *possible* by the fact that the plenum is an indivisible unity; it is rendered highly *plausible* by the fact that the universe, according to the present view, is already a self, animated with a principle of agency, and imbued with something like a will and purpose of its own. In other words, the 'subjectivity' of such a universe is already implicit in its conative nature.[6]

When our candidate for subject status turns out, in this way, to be a cosmic self, an active, global, conative entity, then our conception of the 'subjective' dimension of physical reality, that is, our conception of how that reality 'feels' to itself from within, also undergoes a certain shift. This 'subjective' side of things can no longer be imagined in terms of a passive or static self-presence, but must now be conceptualized as active impulsion. The primal conatus is presumably a vast field of felt impulse, of intrinsic activity, of internal expansions, swellings, dwindlings, contractions, surges, urges, and so on. *Qua* 'subjective' experience, such activity is not to be thought of as occurring *in* space. Rather space, or the order of extension – the plenum – is how this field of inner activity appears externally to observers. In this respect, the 'subjective' or conative field is logically prior to space or extension, since the latter is an order of appearance only. This is not to say that the order of extension is merely illusory. Rather it has something like the status of secondary qualities in Locke's theory of primary and secondary qualities: secondary qualities, such as colour, are, for Locke, grounded in the primary nature of external objects, but there would be no seen-colour if perceivers did not exist. Similarly, in the present case, the order of extension is grounded in the nature of the primal conative field, but there would nevertheless be no seen-extension if perceivers did not exist. What exists, in itself, is this great internally differentiated field of felt impulse. To say this is, again, not to espouse an idealism that, like Leibniz's, renders reality *exclusively* mind-like. The primal field is certainly fundamentally mind-like in nature, in as much as it is a *felt* field, a field of 'subjectivity'; but in other respects it is much less clearly mind-like. This less mind-like character of the primal field is most clearly exemplified in the determinacy, or lawlikeness, of the patterns of its impulses, where such lawlikeness is correlative with physics: an external observer, investigating the order of extension, will indirectly discover laws or patterns pertaining to the nature of the primal field. Thus, while quantum mechanical principles such as wave/particle duality, complementarity and non-locality are descriptive of physical reality, they seem in many respects more applicable to mental than to physical processes, on any traditional account of the latter (Zohar 1990). This ambiguity is less perplexing when such principles are understood as the external manifestation of the mind-like but nevertheless relatively (if only statistically) lawlike nature of the primal conatus.

Thus, the primal field, though fundamentally 'subjective', cannot be characterized in traditional dualistic terms, in as much as it enjoys aspects of both the mental and the physical. In this respect it is perhaps not so different from the notion of energy, which is now arguably the fundamental variable in physics. Energy is pure activity, which exists, or occurs, non-locally, indivisibly and in potentia, in field form, as well as locally, divisibly, and in actuality, in material and other manifest particle forms. Energy is mysterious to physicists precisely in that its many non-classical aspects seem more evocative of mentality than of physicality, as physicality was classically conceived. This mysteriousness dissipates, however, when energy is equated with a primal conatus which is indeed in some sense subjective in nature, though lawlike and manifest as physicality for all that.[7]

If the expression 'Great Impulse' is substituted for 'Great Thought' in Eddington's remark earlier this century that the universe was starting to look more like a Great Thought than a Great Machine, then that remark might be seen as anticipating the present theory. According to this theory, the universe *is* a great, infinitely modulated impulse, and physics is the study of this field of impulsion from the outside – from the vantage point of an observer.[8]

However, to speak of an external observer in the present connection appears, on the face of it, to be self-contradictory. When the object of observation is the universe in its entirety, how can an external observer be posited? More to the point, how can we, as embodied components of the order of extension, qualify as such observers?

While there can be no observers external to the primal field, or reality as a whole, this does not in itself entail that there can be no observers; for the universe itself, under its exterior aspect, is differentiated into local modes, some of which may be capable of experiencing themselves as distinct unities, or centres of subjectivity, separated from the greater whole. Such *secondary* subjects, or centres of subjectivity, would then be capable of functioning as observers of the universe.

To explain what I mean by this, let us consider what would enable a (non-discrete) part of the primal field to become differentiated into something which might justifiably be described as a distinct (though relative) individual. This is a question which I have posed, and answered, elsewhere (Mathews 1991: chapter 3), and my answer was foreshadowed in this chapter by my earlier remarks about self-realizing systems. While I do not have space to detail the relevant argument here, it is, in briefest outline, that wherever the primal field assumes the configurations characteristic of a self-realizing system, it is justifiable to speak of a distinct individual, even while it is understood that such an individual is still ultimately continuous with the primal, indivisible whole. Self-realizing systems, or 'selves', are, as I have explained, systems which, while having the features characteristic of ordinary cybernetic systems – homeostasis, self-regulation, goal-directedness and equifinality – are dedicated to a very special end, namely their own maintenance and self-perpetuation. They are, in other words, reflexive systems.[9] A self-realizing system may thus be thought of as a causal process which, instead of following the usual linear or branching path, loops back on itself to become self-perpetuating. It sets up an enduring, stable unity where before there was only contingent flux. This unity is defined by function rather than by spatiotemporal boundaries or geometric form, the form itself being mapped by the function. Since such systems are self-realizing, they may, as I also mentioned earlier, be attributed with conatus, the impulse towards self-maintenance and self-increase.

The paradigm instances of selfhood, in this sense, are organisms, but ecosystems, and other higher-order systems, could also qualify in principle. Individuation is not, in the case of such systems, precise, and questions of demarcation will certainly arise: there will not necessarily be a clear boundary line between two adjoining ecosystems, for instance. A question also arises as to the status of subsystems within a self-realizing system: do cells, or the kidneys, or the circulatory system, in mammals, for instance, constitute distinct self-realizing systems (Thompson 1990)? This question can largely be answered by pointing to the requirement of self-realizing systems that they be proactive in maintaining their own existence, where this entails procuring their own energy supply: although mammals depend on their native ecosystems for their sustenance, they do actively seek out food and water for themselves, whereas kidneys rely passively on the body as a whole for their nourishment. In general, however, we cannot expect the individuation of self-realizing systems to be absolutely precise. This in no way detracts from the objective though relative functional unity and integrity of such systems.

With their relative individuality, and their conative nature, it seems plausible to assign subject status to these systems – albeit subject status of a secondary or derivative order: in them the 'subjectivity' of the primal field turns in on itself, and becomes locally self-referential. Such systems, or selves, need not be conscious, but if, like us, they happen to be so, they can indeed function as observers. To such observers the primal field will appear as an order of extension, and the excitations within it as physical entities.

The primal Oneness of the world, then, need not preclude the emergence of the relative, self-realizing Many. As members of the Many, we might well wonder how we are to negotiate our own relation to the world in light of this general dialectic between Oneness and Manyness: as conative beings, we are essentially self-interested, but as parts of an indivisible whole, our individual self-interest appears to make little sense. We might also wonder how we, as individuals, are appropriately to relate to a world which is itself a great conative being, a being which is a centre of subjectivity in its own right, as well as a field of secondary subjects. It is to these fundamentally ethical – or perhaps spiritual – questions that I shall turn in the final section.

PART III

Taking up the former of these two questions first, how are we to negotiate the seeming contradiction between the requirements of our individual conatus and the implications of our recognition of the subjective unity of reality as a whole? Should we, in light of our recognition of Oneness, give up our own relative individuality as illusory? Should we resist the imperious promptings of our conatus, and opt instead for a policy of 'no-self'? This is the path generally taken by Buddhists, some of whom – in parts of the Mahayana and Vajrayana traditions (Reynolds 1989; Dowman 1994; Rinpoche 1992) – describe the fundamental nature of reality in terms comparable to those I have used here. Such Buddhists speak of primal mind as the ground of being, and such mind as having a 'sky-like' or space-like character. Individual phenomena arise and pass away in this primal field, while the field itself remains unaffected by their passage (primal mind is the 'mirror'; phenomena are mere reflections in the mirror). Individual phenomena arise or originate codependently, after the fashion of waves in a field. There are thus no individual essences, no truly discrete things, with a nature of their very own. To perceive things as having such a nature, as we ordinarily do, is to succumb to illusion. Enlightenment consists in the casting off of this illusion, the illusion of selfhood – both one's own and that of others. The enlightened being gives up the strivings and resistances that accompany individual conatus. The Many are relinquished in favour of an ineffable, underlying One.[10] In giving up one's individual struggle, and allowing one's self-seeking impulses to arise and pass away harmlessly, 'liberating' themselves at their moment of origin, one is purportedly filled with compassion for all those still caught in the mirage of the Many, still straining to further their own fictitious causes.

Is this how we should resolve the tension between the One and the Many in the present connection? Is the stance of no-self, and the attitude of compassion, the appropriate response to the conundrum? While such a stance does follow with a certain logic from the recognition of Oneness, it is, in my view, weighted too heavily in the unitive direction. To focus exclusively on the Oneness of reality is to ignore the fact that the infinite One also differentiates itself into the finite Many. Why does it do so? Buddhists seem to lack an answer to this question, or at any rate to place no value on the cosmic act or fact of self-differentiation.

The present theory does provide at least the glimmerings of an answer, and this answer derives from the conative nature of the primal field. From the Buddhist point of view – or at least that version of it under consideration here – primal mind is not a self, or entity of any kind, but an unconditioned ground or underlying state; thus, it is not really a 'one', though it does indeed possess the unity and indivisibility characteristic of subjectivity, a unity and indivisibility captured in the metaphor of sky-likeness or spaciousness. Since unconditioned mind in this sense lacks any informing ends or purposes, we cannot ask why it gives rise to conditioned individual phenomena: such mind has no 'reasons'; it is simply in its 'mirror-like' nature to give rise to 'reflections' as it does. From the viewpoint of the present theory, however, the primal field does constitute a 'one', a being which is not only an indivisible unity, but which is endowed with a project of its own, namely its own self-realization. As a conative being, the universe is, from this point of view, pre-eminently creative: conatus is the impulse towards self-creation, and self-creation is surely, logically, the original creative act.

This creativity offers clues to the question of why the One differentiates itself into the Many, and hence to the further question of what value the Many might possess in the larger scheme of things. Perhaps, in order to create itself, the primal One has also to differentiate itself – perhaps its self-actualization hinges on the intrinsic dynamism which is expressed in its differentiation into the Many. Or perhaps, by bringing forth the Many out of its Oneness, the primal subject is serving its own impulse towards self-increase through self-iteration: it conjures a whole new dimension of itself – the world of finite things – out of its pre-existing identity. Such self-differentiation might even be seen as an expression of something like cosmic 'generosity' (indeed 'love'): the primal One creates out of the fabric of its own being the gift of individual existence to bestow upon its resultant creatures.[11] This giving of the self on the part of the One may confer benefits on the One itself as well as blessings upon the Many. For by bringing relatively independent beings to life, the Creator generates possibilities of relationship or interaction for itself, where its interactions with its creatures and their responses to it may in fact expand, or intensify, or renew, primal being.

In any case, there seems to be good reason, from the present point of view, for construing and celebrating the differentiation of the One into the relative Many as an ongoing process of cosmic creativity, rather than spurning it either as a meaningless contingency or as a kind of tragic accident or mistake. Instead of repudiating our selfhood, we might accept both it and the entire order of Creation as expressions of cosmic magnanimity, while also remembering its merely relative or derivative status, and at the same time taking refuge from the undeniable burdens of individual existence in the fact of primal oneness.

If the relative legitimacy of the path of individual self is accepted, then the second question I set out to explore in this section presents itself: how can I follow the promptings of my own conatus while acknowledging that the world is a field of both primal 'subjectivity' and secondary subjects? How can I heed my own elbowing, self-seeking drives in such a world of morally significant others? Can I reconcile the designs I need to make on them for my own conative ends with the consideration that is due to them as subjects, as moral ends-in-themselves?

Perhaps this apparent tension between self and others dissolves when we recall to mind how in the present context (finite, as opposed to cosmic) selfhood is constituted, and what is required for its realization. On the present account, selves, being self-realizing systems, maintain themselves through continuous exchange with things external to them. They are thus essentially relational entities: their ongoing identity and integrity are functions of incessant give and take with elements of their environment. This is so not merely at a material, but at a logical, level: the identity of a (finite) self-realizing system is a function of the identities of those with whom it is interrelated. When selfhood is understood in this way, conatus is served, not by competition, at least in any absolute sense,[12] let alone by a will to stifle and destroy others, but rather by mutually sustaining interaction with them: it is by *mutualistic* relations – relations which promote the flourishing of those who contribute to my own flourishing – that I assert and consolidate my selfhood.

The image of self evoked by this account of self-realization is not one of a triumphant, solipsistic ego, with its boot on a pile of fallen others.

The image is rather one of a self propelled by desire, reaching out to touch and taste others, to contact their inner reality or subjectivity, for the sake of the potentiating, enlivening, transformative effects that this moment of contact, of the sharing of selves, has on its own sense of self. This is thus a portrait not of arrogant domination, dualistic opposition or imperialistic exploitation, but rather of irrepressible *eros*: eros is the appropriate modality for selves which are constituted through mutualistic relations with others who are themselves subjects. It is through eros that such selves nourish their essence as self-realizing beings, that they faithfully serve their conatus. Respect for conatus then does not imply the kind of disregard for others that egoism is taken, in traditional moral philosophies, to imply, where egoism in the latter sense is contrasted with some altruistic notion such as compassion, which subordinates self to others.

So while egoism is the corollory of a view of the Many that negates the One (and is hence the path of the separate, oppositional, 'atomistic' self), and compassion is the corollory of a view of the One that negates the Many (and is hence the path of no-self), eros is, according to the present argument, the path of the relational self, the self constituted in an inter-subjective world in which the One and the Many are in dialectical tension. The 'erotic attitude to reality' (Dinnerstein 1987) is an attitude of openness to the 'subjectivity' of the world, and of playful, life-giving engagement with its intelligence.

But can eros carry the required ethical load in this connection? Does it entrain the kind of regard for the world that the present theory, with its broadscale restoration of 'mentality' to 'physicality', demands? Would not the ideal of compassion provide a surer foundation for our ethical relation to such a reanimated or re-enchanted world?

I think not. While compassion, in its Buddhist and other senses, will surely remain an important thread in our moral attitude to a re-enchanted world, I have reservations about its overall appropriateness in this connection. Compassion involves 'feeling or suffering with' others, but in practice we tend to suffer with others only when we ourselves are not suffering, or not suffering as much as they are. The oppressed themselves are generally too burdened by their own sufferings to take upon themselves the sufferings of others. Thus, in practice, compassion tends to connote pity: we feel compassion for those whom we perceive as in some way worse off than ourselves. In the scenario of spiritual enlightenment, the enlightened one experiences compassion for those who, unlike herself, are still in the thralls of illusion. So, although she has surrendered self, she has also, in another sense, transcended others: she looks down on them from a higher place of detachment, security, invulnerability. For ordinary, fallible mortals, this is a morally perilous pose, proximate to postures of patronization and condescension which are anything but selfless. It is better and morally safer to admit the claims of self from the start, and reconcile those claims with a genuine appreciation of others through eros. Eros involves encounter without patronization; it induces a sharing of self or subjectivity with others, with the result that one cannot remain benignly aloof, as one who assumes the attitude of compassion can. In eros, one 'gets down', gets messy, risks loss of dignity. Compassion is bestowed from a position of self-containment and self-sufficiency; eros seeks and shares from a position of wanting. In short, eros, in acknowledging the needs and desires of self, remains humble, while compassion (paradoxically, given its association with the path of no-self) runs the hidden moral risk of arrogance.

There are other ways in which eros, without being pious, without aspiring to be good, or pursuing the Good itself, nevertheless gets the moral job done. My erotic contact with the subjectivity of other creatures ensures that they become real to me, that I can henceforth never represent them to myself as mere object. Although I will not necessarily sacrifice my own interests to theirs, their interests are now visible and salient to me, and find their way imperceptibly into the muddle of my own. So while I may not, in the name of eros, shoulder the other as a burden, in a spirit of duty or obligation, nor will I ever be able to escape completely the implications of our moment of contact; through inter-subjectivity the other and I have become, however minimally, *wedded*.

Finally, eros is a uniquely affirmative attitude to the world. Provided the other is desired not for the contingencies of their identity (their looks, for instance), but for their subjectivity, their very self, then that desire, that sheer wanting of the other, is surely the highest compliment one self can pay another. The attentions of compassion, on the other hand, are not necessarily flattering, but may be subtly demeaning, implying as they do some lack or condition of disadvantage on the part of the object.

Before closing I would like briefly to consider the question of whether those things which I have here identified as subjects, or centres of subjectivity – where these include self-realizing systems ranging from organisms to the universe at large – are in fact the only loci of subjectivity in the world, and hence whether they are the only things with which we can engage in relations of mutuality, call-and-response, erotic communion. What about non-living parts of Nature? Does it make any sense to invoke the 'spirit' of a place, such as a mountain or a desert spring, or to greet one's native land and expect to be 'recognized' by it,[13] or to harbour a special affinity with a particular object or building, such as the house in which one grew up? The difficulty in all such cases is that there are, in inanimate matter, no natural unities to which one can point as the outward manifestations of inner unities or (secondary) subjects. Places lack determinate spatiotemporal boundaries, and even in the case of objects which do appear to be clearly spatiotemporally defined, such as houses or cars, their boundaries are still ultimately nominal – it is we who choose to regard the material aggregate in question as a whole, or single object; other ways of 'carving [the relevant portion of] reality at the joints' are always conceivable. So, if there are no naturally appointed subjects in the inanimate world, can selected parts of that world be responsive to us?

To answer this question would require an examination of the way the One is present in its subjectively undifferentiated parts. I think that there *is* a sense in which we may 'call up' the One in its inanimate modes, which is to say, in particulars such as places or landforms or even, perhaps, in certain familiar and time-worn artefacts. But explanation of such calling up will have to await another occasion.

NOTES

1 My thanks to Brian Ellis and Hayden Ramsey, and to members of the La Trobe University Philosophy Seminar for comments on earlier drafts of this chapter.

2 If the notion of disposition is, on the other hand, understood in a non-vacuous sense, as implying some kind of real but unobservable causal power inherent in particles, then the implied analysis of what makes things real, as opposed to merely phenomenal, is a variant of the one that I shall outline later in this section (see note 5).

3 Tibetan Buddhists have yogic techniques – 'deep sleep practices' – to enable adepts to remain aware of their self-presence in the sleeping state (Chang 1977).

4 The metaphysics I am developing here contrasts starkly with that of Leibniz in that my ontology is not atomistic, as his is, but is rather holistic, though the primal whole or One on my account also differentiates itself into the Many.

5 Some philosophers of science (such as Brian Ellis, who raised this point with me) would reject my claim that imputing a kind of subjective interiority to matter is the only way of giving content to the appearance/reality distinction. They argue that primary properties such as mass and charge are grounded in causal powers or forces. It is, therefore, implicitly, the presence of such powers or forces in things that renders them real as opposed to illusory or hallucinatory. However, if the notion of causal power or force is to amount to more than the kind of reification of empirical conditionals that I dismissed earlier – if it is to offer a conceptualization of a concrete something underpinning the empirical behaviour of things – then I think that its meaning must ultimately be drawn from our experience of our own subjectivity. That is, the proposition that one thing, X, is *moved by* another thing, Y, has meaning over and above the proposition that X moves after contact with Y, only in as much as the first proposition is derived from our own experience of agency – our experience of making things move. But agency is a function of subjectivity, since it involves not mere motion, but willed or intended motion, where motion can only be willed or intended by a subject. For this reason, then, I think that the argument that what makes the material realm real as opposed to merely phenomenal is that it is a manifestation of causal powers or forces must be seen as a variant of the argument that what makes it real is that it is in some sense subjectively present-to-itself – or it must be seen as such at least until exponents of the causal powers view are able to offer an explication of their

position that does not trade on implicit appeals to our experience of our own agency.

6 I did not draw this implication out in my original arguments for the conative nature of the universe.

7 It might be objected that this lawlikeness would cancel out the free will which is often taken to be definitive of subjective processes. But this does not necessarily follow. The laws in question might relate to probabilities, as in quantum mechanics: a particular impulse in the primal field might occur as the result of the collapse of a wave packet, and as such its actualization might represent just one possibility within a determinate field of possibilities; in this sense its occurrence might fall under laws without being predetermined (see Zohar 1990 for further elaboration).

8 It should by now be evident that this conative view of energy provides a perfect base for the causal powers view of reality that was considered in note 5: the conative view, like the causal powers view, represents all physical processes as ultimately the manifestation of inner impulses, yet it does not suppose that such impulses can be characterized without appeal to some extended or analogical notion of subjectivity.

9 The definitive significance of reflexiveness in this connection is borne out by the etymology of the word 'self', which derives from an Old English term which meant 'the same, the very', and designated a reflexive grammatical function.

10 This should not, however, be understood, in the Buddhist context, as the merging of the finite mind with an infinite or cosmic mind, if 'infinite mind' is understood as an infinite process of thinking and feeling – an infinite unfolding of phenomena. 'Primal mind' in the Buddhist sense denotes the ground of consciousness, pure awareness, or what I have here called subjectivity, abstracted from all its contents: thoughts, feelings, indeed impulses. It is the 'mirror' which makes the 'reflections' possible but remains itself perfectly unchanged by them (Reynolds 1989: 104–5).

11 This reading of the relation between the One and the Many is roughly congruent with the Christian myth of the Crucifixion: if Jesus is symbolic or representative of the incarnation of the One (God) in the finite Many, and if the sufferings of Jesus on the Cross symbolize the travails of mortal life, then the myth demonstrates the price God is prepared to pay to confer the gift of (relative) individual existence on His creatures/modes. ('For God so loved the world that He sent His only begotten son. . . .')

12 This is not to say that predation does not occur, or more generally that members of one species do not consume members of others. However, such inter-species relations are generally, in undisturbed ecosystems, mutually sustaining: that is, the population of the consumed species is generally maintained at optimal levels by the activities of the consumers. The mutualism of ecological relations thus often has to be read at a generic rather than at an individual level.

13 Deborah Bird Rose explains the sensibilities of Aboriginal people of northern Central Australia in this connection: 'Country, or the Dreamings in country, take notice of who is there. Country expects its people to maintain its integrity, and one of the roles of the owners is to introduce strangers to country. Trespass – use of country without permission or introduction – is a threat to the integrity of country . . . A number of people explained that once a person has been introduced to the country . . . the country knows the person's smell. Without this introduction, strangers are at risk – the water may drown them, or they may become sick and die' (Rose 1992: 109).

6 The recovery of wisdom
Gaia theory and environmental policy

Anne Primavesi

Gaia theory, as stated by James Lovelock, focuses on the concept of the Earth as a tightly coupled system where its constituent organisms coevolve with their environment. Self-regulation of climate and of chemical composition are emergent properties of the system, enabling it to respond to stress and perturbation, both natural and anthropogenic. The living and non-living components interact as two tightly coupled forces, each one shaping and affecting the other through systemic feedback loops. The Earth and the life it bears are seen as a system which regulates the temperature and the composition of the Earth's surface and keeps it comfortable for living organisms. This active process of self-regulation is driven by free energy available from sunlight (Lovelock 1991 and 1995).

Gaia theory considers the process of self-regulation in the Earth system over a vast time scale. Lovelock is well aware of the present and probable effects of human activity on that process, especially through pollution emissions and deforestation, and he contributes directly (through his invention of the electron capture and photo-ionization detectors for environmentally hazardous substances and through interdisciplinary research) to international environmental policies intended to deal with those effects. I want to highlight a more indirect contribution. Lovelock's picture of Earth as a coevolving system changes our self-perception by showing us a planetary framework within which we are tightly coupled with our environments. Therefore, it plays a propaedeutic role in environmental policy.

Richard Norgaard's inclusive understanding of coevolutionary theory (as the coevolution of cultural and ecological systems) encompasses beliefs and values, social organization, technology and the environment in coevolutionary process. He posits an ongoing, positive as well as negative, feedback between components of evolving systems, building on the idea that these system components have a variety of traits that are context/environment specific rather than universal, and that they change over time. This then affects the evolution of an environment, where selective forces in turn affect the evolution of each species. The feedback loops give coevolutionary systems the potential for controlled amplification, for producing something new rather than more of the same. These emergent properties also make systems' boundaries infinitely difficult to establish, as parts and relations coevolve and change at varying rates (Norgaard 1994: 28–91; Wilber 1995: 63–6, 532–3). This is as true of social systems as any other, and the complexity of their interactions with ecological systems calls for imagination and flexibility in environmental policy rather than political or economic expediency. Maintaining the distinction between a restrictedly complex function system, such as law or politics, and a much more complex environment is a prerequisite for understanding system/environment interconnections (Luhmann 1986: 12–17; 1989: 6–14).

Such theoretical expansion exemplifies Michael Polanyi's claim that the intellectual passion which informs scientific discovery does not merely affirm the existence of relationships which foreshadow an indeterminate range of future discoveries. Scientific discovery has other effects which Polanyi calls its creative work: creative because it changes the world as we see it by deepening our understanding of it. The change is irrevocable. Having made such a discovery, I shall never see the world, or myself, as I did before (Polanyi 1958: 143).

This change in perspective is an effect of coevolutionary theory especially important for those who make environmental policies. Its importance derives from the processes by which it shifts our vision. How we understand the environment, how we imagine it works to sustain itself and us, should ideally be prior to any policies we make in its regard. Polanyi examines scientific discovery's knock-on effect on our vision by analysing the process of comprehending. It is, he says, a grasping of disjointed parts into a comprehensive whole. The parts are already there, waiting to be discovered. We cannot, he says, comprehend a whole without seeing its parts, but we can see the parts without comprehending the whole. This advance to an understanding of the whole will often require sustained or difficult work, indeed, so difficult that its completion will represent a discovery: the discovery of a particular set of items as parts of a whole. When this happens, the focus of our attention is shifted from the hitherto uncomprehended particulars to the understanding of their joint meaning (Polanyi 1959: 28–9). This sums up Gaia theory's special role in environmental policy.

Polanyi based his analysis of comprehension on the contemporary psychology of gestalt, which he defined as showing that perception is a comprehension of clues in terms of a whole. He disagreed, however, with the then prevailing view that this perception goes on without any deliberate effort on the part of the perceiver and is not open to correction when the results are reconsidered (Polanyi 1958: 97). Psychologists, he said, have described our perception of gestalt as a passive experience; they are unwilling to recognize that knowledge is shaped by the knower's personal activity. He saw this activity as decisive for our understanding of knowledge and, correspondingly, for our appreciation of our position in the universe.

Arne Naess takes this for granted in his use of gestalt, and he makes a significant contribution to the discussion by insisting on the role of sensory as well as intellectual elements in what he calls 'apperceptive gestalt'. He quotes J. Baird Callicott on the ability of ecology to change our values by changing our concepts of the world (and of ourselves in relation to the world) by revealing new relations among objects which, once revealed, stir our ancient centres of moral feeling. Naess comments that the stirring is part of a gestalt and, as such, not to be isolated from the 'objects'. So he, in his turn, supplies, or makes explicit, something only hinted at in Polanyi's use of the word 'passion'. And he takes us further still in understanding the role of our personal knowledge. Our work in unfolding a view of the totality of all that is, is, he says, itself a part, a subordinate gestalt, of that very totality. We are, when active in unfolding our views, creative in shaping and creating 'what there is' at any moment (Naess 1990: 67–80). There is an evolution in our views which is part of the evolution of human knowledge.

What I am suggesting here is that the change in vision brought about through an engagement with Gaia theory marks a phase-change in human understanding of the environment, and that this understanding is essential in the formulation of good environmental policies. It is a comprehension of earth systems in terms of an Earth system.

There is, however, a much-debated postmodernist issue involved in using Polanyi's concept of 'whole' in relation to 'part' and Naess's use of the term 'totality', even though the former's assumption that the 'parts' are 'already' there is in some measure modified by Naess's insistence that our activity shapes and creates what is there. Giving this issue due attention would require, at the very least, a bibliographic essay (Norgaard 1994: 191–4; Soulé 1995: 3–17, 47–65). But I am also conscious that a concept of the earth as a whole is, as far as I know, generally accepted, especially as satellite cameras have fixed the image in our minds, and that its reality, in the way in which that is usually understood, is one on which nearly all of modern science is based.

But that does not excuse us from developing a postmodern ecological sensibility. The theologian Michael Welker describes this as a grasp of the differentiated realities and forms of consciousness created in and by modern society. Liberation theologies and contextual theologies see these differences (religious, economic, cultural, ethnic or environmental) as a call to justice: a call to refrain from colonization in any form. It follows then that theologians, too, must be cautious about any assumption of the 'unity of reality' and of 'the unity of experience'. Instead, sensitive to the complexity and differentiation of systems and environment, we assume a 'reality' or 'realities' that consist of a plurality of structural patterns of life and of interconnected events.

Welker stresses the positive character of postmodern sensibility. It looks, he says, for the resources that counteract the processes of disintegration and destruction, attempting to perceive, conceive and prudently influence processes of emergence. He characterizes emergent change as those conditions and structures whose appearance on the scene cannot be derived simply from preceding conditions and structures, although diverse elements that define both past and present identity persist in them. The past world and the former identity of the elements involved in the process of emergence are also seen anew. And, he adds, there is nothing at all left over (Welker 1994: 28, 37–9). This understanding of emergent process correlates with coevolutionary views of the emergent properties of systems.

As I see it, our sense of the sacred, too, has coevolved in an emergent process which occurs within a relation with the Infinite which, in Emmanuel Levinas's words, exceeds, transcends, overflows the present, the 'what is', by haunting it, disturbing it, stirring the centres of our moral feeling to give testimony to the Infinite, to God. Levinas calls this active testimony our production of infinity: the improbable feat whereby a separated being fixed in its identity, the same, the I, none the less contains in itself what it can neither contain nor receive solely by virtue of its own identity (Levinas 1969: 26–7, 245; 1985: 13, 106–7).

There are many forms of testimony, many elements involved in the emergent process. Welker, working with a theological understanding of the human psyche and its interactions with the community, makes a provocative proposal. Reflecting on Paul's teaching on the work of the Spirit, he names the complex and far reaching effects of this work as 'free self-withdrawal for the benefit of others'. It gives fellow-creatures open space and possibilities of development that surprise and delight them. It does not pin them down, nor does it make claims on them. It finds its most complete expression in love (Welker 1994: 248–9; Luhmann 1986: 174–8).

This proposal points to an open space in environmental policy. The usual questions asked in policy-making focus on environmental impact, on resource-generation, on waste management, on sustainability of ecosystems (Grundelius 1994: 205–6). The questions which are not asked, of the individual and of the community in question, are: What does this ask of me? Who benefits directly from it? Who loses? How and with whom do I cooperate in this process? How do I act in such a way that I can live with what another experiences as a result of my actions? What do I sacrifice?

The word 'sacrifice' shocks our consumerist cultural consciousness. It may also shock evolved religious consciousness, as it carries connotations of bloody rituals, of unhealthy self-negation and/or of helpless victims. In the Christian tradition, interpreting Jesus's death solely in terms of sacrifice for the forgiveness of human sin is now seen by some as creating a religious culture in which the focus on his death rather than on how he lived has allowed, indeed legitimated, victimization through the use of power or wealth. Sacrificial victims used to procure forgiveness, pleasure or peace still abound, both human and non-human (Girard 1977; Condren 1995: 160–89; Gudorf 1992).

So I shall be very specific here. I understand sacrifice as every action done to unite us in sacred community. This keeps the sense of Augustine's classic definition (Augustine 1950: book x, section 6). I recovered this sense when, during a talk on quantum processes, the physicist David Peat used it to describe the transformation of elementary particles. He spoke of the processes of which the electron is a manifestation 'borrowing' a little energy from the universe to make the transformation possible. Thanks to this gift, the electron transforms into a different particle. This new particle only exists for a short time, and as it dies back into the processes that gave birth to it, it pays back the energy it borrowed. So the balance of energy in the universe is maintained (Peat 1996: 84). Referring to both gift and payback as sacrifice, he said that this reciprocal participatory process is the basis of all energy relationships.

Thomas Berry further enhances this understanding of the function of sacrifice within the great community of existence. Comprised of the human community, the life community and the earth community, throughout its evolution all its great transition moments are sacrificial moments. He considers the supernova explosions of the first-generation stars to be an example of such a self-sacrificial moment, enabling everything afterward to come into existence. Sacrifice occurs when something is given and a response is made. Both gift and response cost something. If we are given a physical gift, it is not necessary to give a physical gift in return. We can give the gift of gratitude. (Implicit here in Berry is the understanding of Eucharist as 'The Great Thanksgiving' by and for all that sustains life on earth.) He says that the root of our present tragedy might be considered our unwillingness to make the return. (Or has it something to do with our incapacity to enact meaningful rituals?) Berry sees the entire industrial system as an effort to bypass the return due for our present comforts. Confucius, he says, gave his students a single word to sum up all his teaching: reciprocity; the bonding, the giving, the receiving.

This, he concludes, is why we have sacrifice. I would prefer to say: why a concept of sacrifice is needed in environmental policy. We are bonded to the universe through manifold interactions with our environments, and in spite of the enormous asymmetry between its gifts to us and any response we can make, and in spite of the fact that, in an ultimate sense, there is nothing we can give in return except what we were given, we consciously give something back, even if it is only the conscious acknowledgement of being given something. (This is what I understand Naess to mean when he talks of his 'Gaia gift'.) This process continues until we make our ultimate return in death.

The sacrificial mode itself, according to Berry, can be explained in terms of the different selves that we have: our personal self, our family self, our earth self and our universe self. Ultimately, he says, sacrifice is the choice of the larger self (Berry 1991: 131–7). I agree with him, but would have to add that such a rich self-understanding is not part of conventional Western religious education. It does not give us a sense of ourselves as part of the sacred universe community. Nor is a sense of a greater self part of conventional secular education (although Naess's philosophy offers such an expanded self-perception). It would be dismissed by many environmentalists as unscientific.

It is here that a broader environmental education would benefit policy, enabling us to see ourselves within a personal environment, a family environment, a tribal environment, a cultural environment and a planetary environment. The physicist Freeman Dyson points out that each has evolved on a different time scale, and that each of us is the product of adaptation to the demands of all five time scales. In order to survive, we need to be loyal to ourselves, to our individual impulses and demands. The personal and the central struggle for us is the conflict between that and the demands of the group. This struggle is writ large in environmental policy-making, and will continue. But at the very least, it needs to be met with a conscious and sustained effort to reinforce our larger loyalties. Nature gives us a robust desire to maximize our personal demands. But, Dyson remarks, nature also gave us love in its many varieties (Dyson 1991: 342).

We may or may not be able to love those for whom we make sacrifices, those in whose favour we withdraw. We may or may not see their faces or their forms. Our love and our sacrifice may or may not be reciprocated. But in the emergent processes of our personal environments, the effects of love and sacrifice 'never end'. They 'grow', transform themselves as a force field which both reaches out and draws in, which reshapes and creates diverse webs of relationships while never being reduced to an abstract unity. This constitutes their power in spite of their seeming public insignificance. They can influence judicial, moral and political processes without letting themselves be taken captive by any one of these. In our relationships with particular systems and individuals, sacrifice binds the emotional and the rational into indivisible wholes in which our knowledge of the environment becomes an act of understanding love (Welker 1994: 250f.; Naess 1990: 67).

This stress on active human involvement in creating environments which create human beings is a crucial element in coevolutionary theory. The stress on ethical activity is crucial in the humanizing of environmental policy. By this I mean that we have to factor our own moral evolution into environmental policy: take account of the emergent properties which make us human, such as our capacity for sacrifice and love and our sense of sacred community. Our ability to testify to the Infinite through participation in manifold and complex interactions is a realization of Spirit in the environment. Welker gives an inspiring Christian exposition of the nature of this mystery of sacrifice and love, using the category of 'witness' in ways reminiscent of Levinas's concept of ethical testimony (Welker 1994: 310–11).

The geneticist R. C. Lewontin acknowledges that Darwin's alienation of the organism from the environment was a necessary *first* step in a correct description of the way the forces of nature act on each other. The problem is that it was only a first step, and our thinking has become frozen there. Norgaard argues that in fact Darwin had a more nuanced view of coevolution, but the point here is the way in which, according to Lewontin, Darwin's views of organism and environment have been taken as conclusive rather than indicative (Norgaard 1994: 196). Modern biology, Lewontin says, has become completely committed to the view that organisms are nothing but the battlegrounds between the outside forces and the inside forces, the passive consequences of internal and external activities beyond their control. This view has important political, social and environmental consequences. It implies that the world is outside our control, that we must take it as we find it and do the best we can to find our way through the minefield of life using whatever equipment our genes have provided to get us to the other side in one piece.

What is so extraordinary about this, Lewontin says, is that it is completely in contradiction to what we know about organisms, their behaviour and environment and the rich set of relations between them. Organisms, he says, do not experience environments. They create them out of the bits and pieces of the physical and biological world.[1] Polanyi, Naess, Berry and Welker have indicated some of the complex and manifold interactions effective in creating human environments. In regard to environmental policy-making, this takes us that second step beyond Darwinism to a perception of all our life activities, whether biological, social, imaginative, religious, sensory or political as creative or destructive of environments. The rate and time scale of effect varies enormously, but when, through the lens of Gaia theory, our focus necessarily shifts from the particulars of a specific organism in a specific environment to the understanding of their joint meaning within the climate regulation and chemical composition of the earth system, our vision of those parts changes also. I now become aware of them in terms of the whole on which I have fixed my attention. Polanyi calls this a *subsidiary* awareness of the particulars, in contrast to a *focal* awareness where attention is fixed only on the particulars themselves (Polanyi 1958: 30).

Using Gaia theory to focus attention as widely as possible, we gain a subsidiary awareness which affects our self-perception profoundly. We see ourselves as a particular kind of organism sustained by and sustaining a diversity of environments over a long time scale. This means that those concerned with environmental policy-making cannot approach it on the assumption that there is *an* environment within which our species continues to evolve (or not). Nor can we make policies for the ultimate benefit of living organisms within their environments if we do not see them as the unique consequence of a developmental history that results from the interaction of and determination by internal and external forces: the external ones being what we usually think of as environment. But these forces themselves are partly a consequence of the activities of organisms as they produce and consume the conditions of their own existence. While it will be necessary to focus on an organism or its environment at any one time in policy-making, to deal with one as though it were in reality isolated from the other would be a fundamental error (Lewontin 1993: 63; Lovelock 1991: 25). At the same time, it is the case that our focal awareness will, consciously or not, primarily be attentive to our own needs.

The concept of the 'ecological footprint' of a city strikes a balance between the two kinds of awareness. It quantifies the total area of productive land and water required on a continuous basis to produce all the resources consumed within the city and to assimilate all the wastes produced by its population. The data show that as a result of enormous increases in per capita energy and material consumption, and growing dependencies on trade, the ecological footprints of cities no longer coincide with their locations on the map. Modern high-density settlements necessarily appropriate the ecological output and life-support functions of distant regions all over the world through both commercial trade and natural biogeochemical cycles. So Vancouver uses the productive output of a land area nearly 200 times larger than its political area; London's ecological footprint is 120 times the surface area of the city proper (See Rees 1996, 6–8; Girardet 1992: 86–115; Wackernagel and Rees 1996: 61–124). The anthropogenic stress and perturbation evident around cities can appear to be beyond self-regulation. But perhaps the evolution of the concept of an ecological footprint, emerging from conferences and research around Habitat I and II, brings us a step closer to it.

A deepening awareness of the resources needed to sustain urban communities could also contribute to a subsidiary awareness of sacred community. The recovery of the word and concept 'Gaia' points in this direction. Of Greek origin, Gaia is an early religious name for our planet. Religion, from the Latin *re-ligio*, to re-connect, actively connects or reconnects us with our origins: usually through recounting or re-enacting creation stories in which deities are chief characters. Some structure of meaning and moral guidance based on these stories can be and is appealed to.

From this perspective, 'Gaia' reconnects us with an early stage in our understanding of sacred community. Recounting the creation myth of

Gaia's dance was a religious performative utterance, a speech-act, the whole point of which was to do something for and on behalf of the whole community: in this case, to affirm a belief in the Earth as a living organism and, one may conclude, to live in harmony with this belief. Language, movement and emotion expressed convictions which gave shape to people's lives, and also expressed an intention to act in the future in accordance with those convictions (see Stiver 1996: 80–6, 154–9). Music, movement, symbolic expression and the power of imagination helped to arouse, to articulate and to investigate feelings accepted as guides to right action, and to perceive them in a coherent system that articulated and explained those beliefs to others (Naess 1990: 67). In her study of the myth of Gaia and its evolution, Elisabet Sahtouris insists that the name 'Gaia' was never intended to suggest that the Earth is a female being but simply to affirm the concept of a live earth in contrast to an earth with life upon it (Sahtouris 1989). She gives an informed account of the concept of Gaia as a living planet with a self-creating, self-maintaining physiology which accords with Lovelock's scientific theory.

Gaia became a dormant religious metaphor for the Earth in Western cultural consciousness. But by the time Lovelock used it to name his scientific theory, the myth of an earth-goddess was being revived by forces other than the earth sciences. Among these were researches into the archaeo-mythology of goddess cults in prehistoric Europe and the Ancient Near East, in particular the work of the Lithuanian archaeologist, Marija Gimbutas. Her research coincided with the rise of feminist religious studies and of interest in female deities and their cults. This reawakened interest itself coincided with the emergence of ecofeminism and its radical critique of religious, cultural and social structures for their legitimation of violence against women and nature. At the same time traditional Christianity came under attack for its lack of ecological reference or concern. Within this transformed environment the metaphor of Gaia the earth-goddess awoke to new life: and to his initial astonishment, Lovelock's scientific theory found an unexpected and enthusiastic following.

On reflection, this came as no surprise to Lovelock himself (Lovelock 1995: 191–209). The form and function of religion can be discerned in the way other scientific theories are proposed and received, even though the qualifier 'religious' would not be used or accepted by those who propose or develop them. Lewontin's ironic use of the term 'doctrine' for contemporary sociobiological and genetic theories makes this point. Mary Midgley remarks that what she calls the hunger for unification, once catered for by religious narratives, coupled with the tremendous prestige of science today, leads people to expect spiritual guidance from it as well as technology: an overconfidence now very widely diffused among us irrespective of whether or not we are scientists (Midgley 1995b: 75–84; Midgley 1992b).

The Gaia myth, then, has been revived, to function for some, as it had before, as a religious narrative which reconnects us with our origins. Personifying the planet, talking about it in terms of human physiology (as Lovelock does), it becomes at once both living organism and human environment whose conduct can, in principle, be understood as the reaction of all its parts. The myth has enabled non-scientists to integrate the ever-increasing amounts of scientific data about their environments and their reciprocal relationships with them within an interpretative framework which acknowledges the importance of emotional, aesthetic and religious feeling. I am following Soulé's use of the word 'myth' here, as a conceptual model of nature or specific premises of natural law. The myth may or may not have empirical support and may or may not correspond to observable phenomena but is nevertheless culturally influential, that is, effective (Soulé 1995: 165, n24). It sets our lives in context within the life of the planet while taking account of what we are as a species. It gives us back our personal particularity as human organisms which relate to our planetary environment as active knowers, as creative workers, as understanding lovers or as potential destroyers.

Freeman Dyson sees in the personifying of the planet as Gaia a hopeful sign of sanity in modern society. As humanity moves into the future, he says, and increasingly takes control (*sic*) of its evolution, our first priority must be to preserve our emotional bonds to Gaia. For him, the central complexity of human nature lies in our emotions, not in our intelligence. Intelligence determines means to an end. Emotions determine what our ends shall be. Intelligence belongs to individual human beings. Emotions belong to the group, to the family, to the tribe, to the species. Emotions have a longer history and deeper roots than intelligence. Overcoming the boundary between intelligence and emotion is a function of gestalt formation with profound consequences for environmentalism (Naess 1990: 63). Dyson gives the example of human bonds with the tree. Given the threat of climatic disequilibrium by the accumulation of carbon dioxide in the atmosphere, one which calls for a large-scale international programme of reforestation, he sees the love of trees, implanted in us long ago when we lived on a largely forested planet, as an emotional bond with Gaia which must be preserved as a means of keeping the planet alive. Respect for Gaia, he remarks, is the beginning of wisdom (Dyson 1991: 343–4; Wilson 1992: 349–51).

Our species was classified by Linnaeus in the mid-eighteenth century as that of the 'wise' primates, *homo sapiens*, sapient persons. He quoted Solon's dictum: 'Know Thyself', as 'the first step towards the attainment of true wisdom' (Kerr 1792: 44). The second step before us now would be, I believe, to know ourselves within a Gaia gestalt. Sapiency is defined by Mary Midgley as an understanding of life as a whole, out of which a sense of what really matters in it becomes possible. This power of selection, this knowledge of what matters, is an aspect of wisdom which refuses to seclude one kind of knowledge, science, and cut it off radically from the rest of life. She remarks that Pythagoras forbade people to call him *sophos*, wise, explaining that he was only *philosophos*, a lover of wisdom. Wisdom which is loved and truly valued is seen to be something difficult, which it will take time to search for. What disturbed people about Socrates, she says, was the apparent discrepancy between a genuine, longstanding dedication to facing the large questions and his refusal to produce ready-made answers. This policy, he said, was due to his awareness of his own ignorance, and his only claim to be at all wise rested on that awareness (Midgley 1989: 13, 45, 74–89, 96–7).

Two of his most illustrious followers, Plato and Aristotle, went on to discuss the nature of wisdom in ways which, I hope, point us in the direction of the wisdom needed to make and to implement good environmental policy. Wisdom for the Socratics refers to different aspects of moral, intellectual, aesthetic and social life. There is wisdom as *sophia*, the special gift of the philosopher and of those who have devoted themselves to a contemplative life in pursuit of truth. This is accompanied by emotions such as awe and love, by feelings of wonder at the mystery disclosed in contemplation. The scientific aspect, *episteme*, is a form of knowledge developed by those who study the nature of objects and have the ability to demonstrate or to prove what they are by nature. These aspects are not mutually exclusive. On the contrary. The contemplative aspect, *sophia*, is seen as the union of intuitive reason (whereby we grasp a universal principle or truth in subsidiary awareness) with *episteme*, rational knowledge (focal awareness). The particular aspect of knowledge is defined and dignified by its object rather than its process, which, since it is innate in the human person, is essentially indivisible. We may differentiate between forms of knowledge for clarity, or for convenience, but not in reality.

The other aspect of wisdom is what Aristotle calls *phronesis*, or practical wisdom, that of the statesman or lawgiver who, through consultation, finds the prudent course of action and implements it with regard for the good of the whole community. This participatory form of wisdom is especially relevant to effective environmental policy-making. It, too, is subdivided according to its object. In regard to the individual, it is *phronesis* in the narrow sense. Concerned with the family or household, it is *oikonomia*, economics (cognate with ecology). As regards the state, it is political science in its widest sense (Robinson 1990: 13–23; Aristotle 1986: 146–73).

All of these types of wisdom are needed for a meaningful grasp of the complex systems and interactions which structure the earth community. While *sophia*, contemplation, and *episteme*, science, are still notionally related, *phronesis*, participatory, practical and/or political wisdom has apparently lost its character and function of integrating their insights with what it is normatively possible to achieve within society. This latter form, however, assumes that to be wise is to be a certain kind of person, living in conformity with one's deepest convictions, with one's passions and desires disposed in such a way that one's deliberate choices promote the flourishing of one's human and humanizing attributes, the well-being of one's family and community. The contemplative life is not divorced from this environment, assigned to a 'religious' milieu. Rather it deepens and increases understanding of complex interactions and relationships within the great community of existence: those closest in space and time as well as those to whom we may not consciously advert. Sustaining these relationships through love and sacrifice testifies to Spirit in all environments.

This integration of wisdom and action is systematized in Naess's philosophy, which he names 'ecosophy': a combination of *oikos* and *sophia*, 'household' and 'wisdom'. 'Eco' here has an appreciably broader meaning than the immediate family, household or community. 'Earth household' is nearer the mark. He sees the emergence of human ecological consciousness as a stage in which a life form has developed on earth which is capable of understanding and appreciating its relations with all other life forms and to the Earth as a whole. This conscious change of attitude is a presupposition in all good environmental policy-making, which is concerned with relationships between entities as an essential component of what these entities are in themselves. In ecosophy, the uniqueness of *homo sapiens* is not a premise for relationships of domination and mistreatment, but for a universal care that other species can neither afford nor understand (Naess 1990: 36–7, 163–212).

A coevolved wisdom integrates these aspects of knowledge in an apperceptive Gaia gestalt. It does so by consciously affirming the role of feelings in motivating us to plan wisely and to act accordingly. Emotions bind us to the different environmental communities to which we belong. We give and receive within the Earth household because we love and feel loved, because our compassion is aroused or our sense of awe awakened. We sacrifice ourselves in reciprocal relationships in which morality and reality converge; in which what we do and what we are act as transformative factors in whatever environment we share. Realizing that I cannot encompass all aspects of Wisdom, I respect and acknowledge her gifts in others. Recovering her Spirit, I may go beyond boundaries set by self-interest, acting out of understanding love while sharing knowledge of hope.

NOTE

1 Lewontin 1993: 108–9. Lynn Margulis gives a prototypical example from the Archaean period (which ran from about 3.7 to 2.5 billion years ago) between bacteria (forebears of the mitochondria of our own cells) and oxygen, toxic

to them, released into their environment (Margulis and Sagan 1995: 105–17). See also Briggs and Peat (1989: 155–8) for an account of Margulis's work as a revolutionary feedback theory of evolution as symbiosis: that is, not as brutal competition but as cooperation. Margulis is one of Lovelock's closest collaborators (see Lovelock 1995: 101–6).

7 Nature and the environment in indigenous and traditional cultures

Kay Milton

Participants in environmental discourse often express the belief that non-industrial societies, usually described as 'indigenous' or 'traditional',[1] have a better relationship with their environment than industrial societies do. A growing body of anthropological literature (Ellen 1986, 1993; Brightman 1987; Milton 1996) has thrown doubt on this belief, and suggested that it should be seen as a 'myth', both in the popular sense of something that is untrue, and in the sense often used by anthropologists, as something that is asserted as a dogma (Robinson 1968) regardless of its truth or falsehood. Confronting the myth is not a simple task, however, for it exists in several versions which need to be examined through a range of arguments and types of data. In this chapter I address just one version of the myth: the belief that indigenous and traditional peoples have a 'oneness' with nature. I use this as a starting point for discussing recent work by anthropologists on the concept of nature in indigenous and traditional cultures. This discussion will demonstrate how our understanding of these cultures is shaped by our preconceptions and by the arguments we wish to pursue. It will show that both the reality of human–environment relations, and our interpretations of them, are considerably more complex than the environmentalist myth suggests. The initial task, however, is to distinguish the environmentalist belief about oneness with nature from other versions of the myth, in order to identify the most appropriate way of addressing it. I therefore begin by considering the myth as a whole.

UNRAVELLING THE MYTH

Environmental discourse is peppered with statements about the way indigenous and traditional societies relate to their environments. The report of the World Commission on Environment and Development (the 'Brundtland Report') referred to 'the harmony with nature and the environmental awareness characteristic of the traditional way of life' (WCED 1987: 115). Margaret Thatcher, the British Prime Minister, praised 'the primitive tribes of the rain forests for having a "one-ness" with their environment that has been lost in the urban jungle' (*The Times*, 15 September 1989, quoted in Rayner 1989). Christopher Manes contrasted the megalomania of industrial society with cultures 'that have developed a sustainable and harmonious relationship with their surroundings', naming the Mbuti, the Penan and the !Kung as examples. 'Out of some hidden source of wisdom', he wrote, 'these societies *chose* not to dominate nature' (Manes 1990: 28, emphasis given). *Caring for the Earth*, a 'strategy for sustainable living' published by some of the leading international environmental NGOs,[2] expressed the view that traditional subsistence activities 'reinforce spiritual values, an ethic of sharing, and a commitment to stewardship of the land' (IUCN *et al*. 1991: 61). In his opening address to the United Nations Conference on Environment and Development in 1992 (the Rio Earth Summit), Maurice Strong said, 'We must reinstate in our lives the ethic of love and respect for the Earth which traditional peoples have retained as central to their value systems' (United Nations 1993b: 50). The *International Treaty between NGOs and Indigenous Peoples* states, 'For centuries the Indigenous Peoples have had an intimate relationship with nature, passing along respect, interdependence and equilibrium. For this reason, these people have developed economic, social and cultural models that respect nature without destroying it.'[3]

At first glance, these and other similar statements might all appear to be expressing the same simple conviction, but a closer examination reveals several different ideas. The most important distinction to be drawn is between beliefs about the way indigenous and traditional peoples act towards their environment (that they do not dominate or destroy it, that they live sustainably within it), and beliefs about what they think, feel or know about their environment (that they respect it and possess ecological wisdom). This distinction is important because the relationship between ideology and action is not simple. People who behave in non-destructive ways that enable them to live sustainably do not necessarily respect their environment. Their material requirements may be such that they simply do not need to stretch their environment's capacity to support them. Conversely, people may respect their environment but still act in ways that damage or destroy it (see van den Breemer 1992). They may regret such damage but see it as beyond their control; they may, for instance, regard protection of the environment as the responsibility of a central government or a divine power, rather than themselves. Thus, to say that a community has an ethic or an ideology of respect for their environment is different from saying that they act in environmentally benign ways. This distinction between ideology and action corresponds to that which anthropologists habitually make between 'culture' (consisting of people's thoughts, feelings and knowledge) and social organization (consisting of individual actions and observable patterns of social activity). In anthropology this distinction is made precisely because the relationship between these two spheres is recognized as problematic and needs to be left open to investigation.

We can also distinguish different versions of the environmentalist myth within each of these two spheres. At the level of action, to say that indigenous and traditional peoples do not destroy their environment is not the same as saying that they do not dominate it, since it is clearly possible to dominate without destroying. Similarly (and without going into the definitions of these terms), I suggest that ecological sustainability, harmony or balance is not, in theory, dependent on the absence of domination. Indeed, many environmentalists within industrial society aim to achieve sustainability precisely through domination of the environment, usually described in terms of 'management' or 'development' (see Milton 1996).

At the level of ideology, it is again possible to discern several versions of the environmentalist myth. The belief that indigenous and traditional peoples possess ecological wisdom is different from the belief that they possess environmental awareness. 'Wisdom' is often assumed to mean knowledge, that people know how to treat their environment in ways that do not damage it. 'Awareness' often implies sensitivity, that people are conscious of the needs of their environment. Clearly, it is possible to have an awareness of environmental needs without possessing the knowledge to meet those needs. Again, this seems to be the situation in which many Western environmentalists find themselves. Similarly, the belief that certain peoples 'respect' or 'love' their environment is different from the belief that they experience a 'oneness' with it. Indeed, it could be argued that these versions of the myth are logically incompatible, for in order to love or respect the environment, people must first see themselves as distinct from it, thus denying a sense of oneness. This point will be revisited below.

The number of different versions of the myth multiplies even further when we acknowledge the various labels applied to what I have so far called 'the environment'. In the instances quoted above, terms such as 'the Earth', 'the land' and 'nature' are also used. Within environmental discourse, the meanings of these terms frequently overlap, but they are also often assumed to be different. Thus 'the environment' may be congruent with 'the Earth' when global issues are being discussed, but 'the environment' is also used at the local level, to refer to a community's immediate surroundings. 'Nature' is perhaps the most ambiguous term, referring sometimes to that part of the environment that is separate from human activity, sometimes to an all-encompassing scheme of things that includes human beings along with everything else, and sometimes to a quality of human and other beings that is assumed to be innate rather than acquired.

A oneness with nature

What does it mean to say that indigenous and traditional peoples have a oneness with nature? The statement could refer either to action or to ideology. With reference to action, it could mean that a society, from the viewpoint of an outside observer, appears to be at one with nature in the way its members treat their environment, that their actions are in sympathy with natural processes. In this sense, the statement is making no assumptions about what is in other people's minds, and in order to understand it, we need to know what 'nature' means only in the mind of the observer.

In this chapter, I am concerned with ideology rather than action; with how the above statement might apply to the *cultures* of indigenous and traditional peoples rather than their modes of behaviour. In this case, assumptions are being made about what exists in the minds of such people. In stating that they have a oneness with nature, we are saying that this is how they see themselves, that a oneness with nature is part of their experience, and implying that they hold in their minds, as part of their culture, something that might legitimately be called 'nature'. Thus, in order to address this version of the environmentalist myth, we need to ask what nature might mean to indigenous and traditional peoples, and the obvious way of tackling this question is to consult the anthropological literature on the subject. This task may be enormous but not, on the face of it, conceptually problematic. However, we need to take into account the process of ethnographic description. Anthropologists' accounts of indigenous and traditional cultures can never be 'neutral'; they are inevitably filtered through their own analytical frameworks.[4] So we also need to ask what 'nature' means in the minds of the anthropologists whose work we are consulting. The analysis in the following sections thus takes the form of a conversation between ethnographic 'fact' and anthropological theory.

I suggest that a oneness with nature might be defined in both a weak and a strong sense. In the weak sense, the people in question recognize something that can be glossed as 'nature', and see themselves as distinct from it. They regard the relationship between themselves and nature as harmonious and may describe that relationship in terms of concepts such as 'respect', 'caring' and 'sharing'. The strong sense is that referred to above: the people in question do not recognize a distinction between themselves and their environment; they and the things with which they interact belong to the same continuous system. They recognize relationships with particular human and non-human entities that inhabit that system – plants, animals, spirits, gods, kin, affines, strangers – but not with nature as such. In this instance, the sense of oneness is so complete that it is simply taken for granted; it remains unspoken and unthought of.

THE NATURE–CULTURE DEBATE

These two senses in which indigenous and traditional peoples might be said to experience a oneness with nature correspond, more or less, to current positions in the longstanding anthropological debate about nature and culture. What is at issue in this debate is precisely the question of whether the concepts 'nature' and 'culture' are universal in human thought, and therefore present in all cultures, or whether they have evolved out of certain types of human–environment relationship, characteristic of some, but not all, societies. The issue is complicated by several factors. First, it is widely recognized by anthropologists that, in order for a concept to be present in a culture, it does not have to carry a label; it may be 'terminologically covert' (Ellen 1996: 107), and expressed through actions rather than words (Turner 1967: 20*ff*). Thus, searching for concepts of nature is not simply a matter of looking for words which can be translated as 'nature'. Second, as any anthropologist will confirm, the things people take for granted in their everyday lives are usually the hardest parts of their culture to identify, precisely because they are not articulated in words and may be difficult to infer even from action. So if people do experience a total oneness with nature, in the strong sense outlined above, this might easily be missed by an observer. Third, translation between cultures has always been beset by difficulties. These difficulties have been seen as particularly severe during the past thirty years or so, when 'cultural relativism' and 'cultural constructivism' have been the dominant doctrines in anthropology.[5] But whatever approach is taken, it is difficult to judge the comparability of concepts that have evolved in different social contexts. In the following analysis, I use selected arguments in the nature–culture debate to demonstrate the problems involved in determining whether, and in what sense, indigenous and traditional peoples experience a oneness with nature.

Hunter-gatherers and the absence of 'nature'

One of the most thought-provoking studies in the current debate is Bird-David's analysis of hunter-gatherer cultures (1990, 1992, 1993). It is now widely accepted that hunter-gatherer economies are not clearly distinguishable from others in terms of their mode of subsistence. Not only do some hunter-gatherers engage in other economic activities, including trade and paid employment, but the process of securing a living from so-called 'wild' sources entails a degree of intervention which, in some instances, might be described as 'management' (Schmink *et al.* 1992). Bird-David suggested that hunter-gatherer societies are distinctive, not in their mode of subsistence, but in the manner in which they understand their environment and their relationship with it. Drawing on her own fieldwork among the Nayaka of Southern India, and comparative material on the Mbuti of Zaïre and the Batek of Malaysia, she argued that hunter-gatherers see their environment as a 'giving' environment (Bird-David 1990). These societies interact with non-human agencies as members of a sharing community. The spirits of their forest environments join them to eat, sing and dance (Bird-David 1992: 29).

The people expect to share what the forest has to give, just as they expect to share with their human companions. So striking is the impression of a single, sharing community encompassing human and non-human agencies that Bird-David commented, 'they do not inscribe into the nature of things a division between the natural agencies and themselves as we do with our "nature:culture" dichotomy. They view their world as an integrated entity' (Bird-David 1992: 29–30). This looks like a description of the stronger sense of oneness with nature, outlined above.

Bird-David's account of the giving environment is superficially similar to Ingold's analysis of hunter-gatherer cultures. For Ingold, the relationship between hunter-gatherers and their environment is based on trust:

> The essence of trust is a peculiar combination of *autonomy* and *dependency*. To trust someone is to act with that person in mind, in the hope and expectation that she will do likewise – responding in ways favourable to you – so long as you do nothing to curb her autonomy to act otherwise.
>
> (Ingold 1994: 13; emphasis given)

Thus, 'the animals in the environment of the hunter do not simply go their own way, but are supposed to act with the hunter in mind' (ibid.: 14). Similarly, the hunter acts with the animals in mind. This mutual trust is particularly well documented for indigenous North American cultures. The Cree hunters of Quebec see themselves as belonging to the same moral community as the animals they hunt. They are aware of the needs and sensibilities of the animals and take them into account in their activities. Similarly, the animals are aware of and provide for the needs of humans. Gifts are exchanged between them. The animals give themselves as food to the Cree; offerings of tobacco are made in return, and gifts of food placed in the fire are carried back to the bush via the smoke ascending through the chimney (Tanner 1979: 172–4). Individual hunters are helped by animal friends, with whom they form long-term relationships (ibid.: 139–40).

For both Ingold and Bird-David, hunter-gatherers live in a mixed community of human and non-human persons who mutually respect and share with one another. However, as Ingold himself has pointed out (1996), there is a significant difference between their two analytical images. For although Bird-David stated that hunter-gatherers view their world as 'an integrated entity', with no division between the human and the non-human, her analysis implicitly denies this. For Bird-David, the hunter-gatherers' view of the giving environment is a metaphor derived from human society. Their understanding of interpersonal sharing among themselves represents to them their relationship with the giving environment, and different kinds of giving are expressed through different metaphors drawn from human relationships (Bird-David 1990: 194). The Nayaka, the Mbuti and the Batek see the forest as their parent, unconditionally providing what they, its children, need, while indigenous North American hunters speak of their prey as sexual partners, who need to be persuaded, seduced, into giving themselves (Bird-David 1993).

The concept of metaphor depends on a distinction between spheres of reality which are assumed to be understood in different ways; something has to provide the metaphors through which something else is understood. In Bird-David's analysis, human society provides the metaphors through which human–environment relationships are understood; these relationships appear in hunter-gatherer ideology as symbolized or modelled reality, with human society providing the building blocks for the models (see Milton 1996: 214–15). It would not be possible to describe a culture in this way without assuming, if only implicitly, that it contains a fundamental division between the two spheres, human and non-human (Ingold 1996: 124–5). Thus, while Bird-David's analysis looks like a description of oneness with nature in its strong sense, it is in fact closer to the weak sense. Hunter-gatherers distinguish themselves from their environment and see their relationship with it as harmonious, based on respect and consideration.

Ingold, on the other hand, has argued unambiguously that hunter-gatherers recognize no division between themselves and their environment:

> In their account ... there are not two worlds, of nature and society, but just one, saturated with personal powers and embracing humans, the animals and plants on which they depend, and the features of the landscape in which they live and move.
>
> (Ingold 1996: 128)

In his analysis, the relations between hunter-gatherers and the non-human agencies in their environment are not *modelled* on human social relations, they are relations of the same kind, constituted in the same way, through a process of continued active engagement (ibid.: 125). I am suggesting here that this might be seen as a oneness with nature in its strong sense, but there is a contradiction in this suggestion, for in a description of the hunter-gatherers' world as a continuous and fully integrated entity, there appears to be no place at all for 'nature'; it is not recognizable as a distinct sphere. The strong sense of oneness dissolves the boundaries on which the weak sense depends.

The argument that a concept of nature has no place in a world that is viewed as an integrated whole is strongly supported by Dwyer's comparison of societies in Papua New Guinea (Dwyer 1996) whose world views are characterized by different degrees of integration. A fully integrated world is experienced by the Kubo-speaking community of Gwaimasi. Here, the human population is sparse, their current and former gardens occupy a tiny proportion of the land, and they draw extensively on the resources of their environment, combining cultivation with sago extraction, hunting, gathering and fishing. Through these activities, they range widely over their landscape, endowing it with memories, gleaning from it both material and spiritual sustenance. Thus they live (Ingold would say 'dwell') within their whole environment; it is entirely familiar to them, saturated with significance. The non-material world[6] of spirits and other 'fabulous beings' also occupies the whole environment, interacting freely and directly with humans and animals (Dwyer 1996: 168).

The Siane-speaking people of Leu village, also studied by Dwyer, occupy a very different world. Their population density is much higher than that of the Kubo. They garden intensively and on long-term sites, obtaining 90 per cent of their food from this source. Activities which take them out of occupied areas are relatively rare and take up a small proportion of their time. Their non-material world does not interact easily with the material domain:

> The beings that inhabit it often assume forms that have no parallel in the visible world. They may reside far beyond the accessible domain of people or even in undifferentiated space. In the sense of both form and location they are peripheral beings, who communicate with the living through intermediaries.
>
> (Dwyer 1996: 175)

On the basis of these and a third, intermediate, example, Dwyer generalized about the capacity of human societies to develop a concept of 'nature'. In the fully integrated world of the Kubo, there is no sphere sufficiently distinct from the human world to merit the label 'nature'. Everything occurs 'within the landscape of human action, for there is no other place' (Dwyer 1996: 168). For the Siane, the landscape of human action does not occupy the whole environment, but leaves room, around its margins, for something that might be called 'nature'. As Ingold expressed it, 'the world can only be "nature" for a being that does not belong there' (Ingold 1996: 117), so in order to perceive nature, people must recognize a part of their environment where they feel they do not belong.

Reinstating 'nature'

The analyses presented by Ingold and Dwyer might lead us to ask what we should call the fully integrated world in which some societies live. We cannot call it 'nature', for nature, we are told, can exist only in opposition to the human sphere. Dwyer suggested that we call it 'culture', that because the whole environment of the Kubo is endowed with significance for them, it is a cultural rather than a natural landscape (Dwyer 1996: 178). But if nature and culture are defined, for the purposes of the analysis, in opposition to each other, then an integrated world in which no such division exists is no more cultural than it is natural. I suggest that the integrated world in which the Kubo live conforms very closely to one of the ways in which nature is commonly understood in Western society: as the all-encompassing scheme to which human beings and all other things belong. It could be argued that, in treating nature only as a sphere opposed to culture, Ingold and Dwyer have misrepresented the Western concept of nature; they have narrowed it down to just one of its major constituents. As indicated above, the Western concept of nature is ambiguous; it has multiple meanings which shift from one context to another. Ellen has suggested that we take these multiple meanings as a starting point for cross-cultural analysis. By starting from a broad analytical concept of nature we can explore the variation in people's understanding of their environment, without having to invent a new terminology.

Ellen proposed three definitions of nature, all of which are present in the Western concept: nature as a category of 'things', nature as space which is not human, and nature as inner essence. These 'dimensions' or 'axes' can be used as a guide to identifying and describing concepts of 'nature' in non-Western cultures:

> To the extent that these three cognitive axes make an equal contribution to representations, they may be predicted to approach that multi-faceted, ambiguous, but ultimately recognisable idea which we in the West recognise as nature; whereas the more asymmetry is introduced into the model the less familiar the construction becomes.
>
> (Ellen 1996: 105)

In order to define or generate a cultural concept of nature, at least two of the three dimensions are necessary. These are juxtaposed in people's experience of nature (Ellen 1996: 112), leading, quite often, to the kinds of ambiguity and contradiction which characterize the Western concept. For instance, in terms of nature as a category of things, we frequently place ourselves within nature on the understanding that we possess the physical characteristics of animals, and we also speak of ourselves as possessing an inner essence, a 'human nature', which is not derived from culture; but when nature is understood as a space which is not human, we are explicitly excluded from it.

Using Ellen's three definitions of nature, we can see that Ingold and Dwyer consider only the second one. They describe cultures in which 'space which is not human' is not recognized, and conclude, on this basis, that it is inappropriate to attribute a concept of nature to such cultures. And yet, it would be difficult to argue that the other two definitions of nature cannot be recognized in the cultures they describe. Certain aspects of these cultures appear closely comparable with some of Ellen's material on the Nuaulu people of Seram. He suggested that a conflation between the first and the third definitions of nature (as things and as inner essence) is exemplified in the attribution of essence to parts of nature, and that, in Nuaulu culture, this is expressed in the consubstantiality of spirits and animals:

> Nuaulu recognise spirit categories in much the same way as they recognise categories of animal; indeed, spirits are treated as natural kinds, as equally significant parts of their environment.... People claim to hear and 'see' spirits all the time and I have on occasions been present when the alleged discovery of a particular spirit in a tree or a bush has created scenes of some excitement.
>
> (Ellen 1996: 114)

Spirits can enter the bodies of animals, and some animals, insects in particular, appear to be identified as spirits, while others are thought to be derived from spirits. So, in the Nuaulu understanding of nature, there are 'sometimes no simple breaks at domain boundaries in what we might construe as the "real" world', and 'even areas of overlap between the objectively visible and invisible' (ibid.: 115). The parallels with Dwyer's description of the Kubo world view are quite clear:

> The invisible world permeates the land. Fabulous beings are associated with specific environmental zones or even particular places.... The spiritual essences of animals are unconfined by the habitats of their mundane forms and, in the form of animals, the spirits of the dead may be seen in both expected and anomalous places.
>
> (Dwyer 1996: 168)

The world views of indigenous North American societies also appear to exemplify the conflation of Ellen's first and third definitions of nature. The manner in which these cultures attribute essence to natural things was summarized by Callicott as follows:

> the typical traditional American Indian attitude was to regard all features of the environment as enspirited. These entities possessed a consciousness, reason, and volition, no less intense and complete than a human being's. The Earth itself, the sky, the winds, rocks, streams, trees, insects, birds, and all other animals therefore had personalities and were thus as fully persons as other [sic] human beings.
>
> (Callicott 1982: 305)

If we adopt Ellen's broad and multifaceted concept of nature, it becomes impossible to argue that 'nature' cannot be recognized in the cultures of hunter-gatherers. We might say that it is not present in all its aspects, that it is not recognizable as a sphere separated from culture, but we cannot say that it is not present at all. Ellen's approach, by virtue of its broad analytical concept of nature, thus produces a different mode of classifying people's understandings of their environments. Instead of referring to societies which have concepts of nature and those that do not, we can speak of different ways of thinking about nature. Ellen distinguished between those that express continuity between the human and non-human worlds, and thus represent nature in a non-oppositional way, and those that effectively oppose nature and culture (Ellen forthcoming). Separation and opposition are not the same, however. In addition to those ways of understanding that see the world as fully integrated, and therefore express a strong sense of oneness with nature, there are those which separate culture from nature but see the two spheres as operating in harmony (the weak sense of oneness), as well as those which see them in opposition to one another. As Ellen has demonstrated for the Nuaulu, and as we know from Western culture, these variations in environmental understanding can be held within the same society or community. In the final section of this chapter, I consider briefly how such diverse understandings of our environment arise, and what they can teach us about the environmentalist myth.

THE FOUNDATIONS OF ENVIRONMENTAL UNDERSTANDING

However broad or narrow their analytical parameters, anthropologists generally agree that the way people understand their environment derives from their use of it, how they live within it. Several theories have been proposed to explain how specific environmental perspectives arise. As we have seen, Dwyer argued that a non-oppositional perspective is consistent with an extensive pattern of economic activity which makes people familiar with every part of their environment (Dwyer 1996). As Ellen (forthcoming) expresses it, 'Moving around in forest is not conducive, after all, to developing an enduring opposition with it.' Oppositional perspectives, according to Dwyer, develop as people concentrate and intensify their economic activities, creating, in this process, spaces that are not human.

Ingold focused on a similar transformation in the relationship between human and non-human animals. He observed that, while subsistence hunters engage with their prey on the basis of mutual trust, pastoralists dominate the animals on which they depend; they control their movements, care for them and decide when and how their lives should end (Ingold 1994: 16). Control makes trust redundant; indeed, it removes the autonomy which is part of the essence of trust. The non-human animals, once deprived of their freedom, come to be seen as lacking the capacity to act on their own behalf; they are seen less as persons and more as objects. It is easy to perceive, in the intensive farming methods of the industrial world, an extreme form of this perspective, in which the sensibilities of non-human animals are denied for the convenience of human routines and industrial systems.

Coursey commented on the ideological consequences of different forms of plant cultivation. Vegetative cultivation, which involves propagation by roots, tubers and cuttings, generates a non-interventionist attitude to nature, in which humanity is seen 'as an integral part of the overall ecosystem rather than as something above, separate, and dominating it' (Coursey 1978: 139). Seed cultivation, on the other hand, requires a more interventionist approach, in which human activities and natural processes are sharply contrasted, generating an understanding of nature as something separate from and opposed to humanity. Thus, if we were trying to identify an event in the history of Western cultures which fostered the development of oppositional concepts of nature, we should look to the Neolithic revolution, the cultivation of grain crops in the Middle East around ten thousand years ago, and its subsequent spread across Europe. The Judaeo-Christian tradition, often identified by environmentalists as the source of alienation from nature, helped to rationalize an oppositional model which was already present, by providing people with 'a theoretical basis for regarding themselves as a separate and inherently superior creation from the rest of the biological world' (Coursey 1978: 140).

In order to understand how ways of understanding nature come to be distributed among human populations, it is important not to be dictated by the conventional classification of human economies (or ecologies) as hunter-gatherer, pastoralist, shifting cultivator, settled agriculturalist, industrialist, and so on. These labels are not sufficiently sensitive to represent the diverse ways in which people engage with their environment. The activities themselves, the practical experiences of interacting with parts of our environment, generate our attitude towards it and our thoughts about it. Where several modes of engagement are practised within the same community, or where modes of engagement change rapidly over time, we can expect this to be reflected in their ideology. Ellen has shown, for instance, how Nuaulu concepts of nature have been repeatedly renegotiated as their physical environment and their ways of using it have changed (Ellen forthcoming). We can expect contradiction and ambiguity to be particularly prominent in the world views of people who practise hunting and gathering, which tend to produce non-oppositional perspectives on the environment, alongside seed cultivation, which tends to generate oppositional perspectives. Contradiction and ambiguity are nowhere more evident than in the Western tradition, which places humanity sometimes within and sometimes outside nature, and which sometimes reveres nature and sometimes despises it.

CONCLUSIONS

My purpose in presenting the anthropological perspectives discussed above is not to pass judgement. I am not suggesting that anthropologists should adopt either broad or narrow definitions of nature; I am simply pointing out that our choice of one or the other has consequences. Our understanding of what indigenous and traditional cultures are like depends on where we start from. As with anthropologists, so with environmentalists. We should not be surprised if those who see industrial development as the major cause of environmental problems assume, naïvely, that non-industrial societies hold the key to their solution. Nor should we be surprised if those who see alienation from nature as the result of particular events in the history of Western society – industrialization, the adoption of the Judaeo-Christian tradition – see in non-Western cultures the sense of 'oneness' with nature which they themselves have supposedly lost. But much of what anthropologists have learned about the non-industrial world indicates that these views are misconstrued. It is unrealistic to lump together all indigenous and traditional peoples and claim that they understand their environments in ways that contrast sharply with Western models. Indigenous and traditional societies embrace a wide range of ecological practices, which generate a diversity of environmental perspectives, some of which are as ambiguous and contradictory as Western concepts of 'nature'.

Nor are these societies locked into some kind of static relationship with their environments which the West has left behind, as some versions of the environmentalist myth suggest. For many, environmental, economic and cultural changes have been as dramatic as in any industrial society (see Ellen forthcoming). An interesting recent development, encouraged by the process of international and transnational negotiation that surrounded the Rio Earth Summit, is the participation of indigenous and traditional peoples themselves in the environmental discourse, and their adoption of the environmentalist myth as a way of representing themselves in the global arena (Milton 1996: 202). But whatever its role in contemporary discourse, the environmentalist myth can only obscure the reality of human–environment relations, and hinder the sensitive understanding of the relationship between culture and ecology which we need to guide our search for sustainable ways of living.

ACKNOWLEDGEMENT

I am grateful to Roy Ellen for providing both ideas and pre-publication material during the preparation of this chapter.

NOTES

1 The terms 'indigenous' and 'traditional' are not entirely appropriate, and are used here simply because they are the labels most often used in environmental discourse to describe societies whose economies have never been industrial in character. *Agenda 21*, the most comprehensive of the agreements to have emerged from the Rio Earth Summit, describes 'indigenous people' as having 'an historical relationship with their lands' and as being 'descendants of the original inhabitants of such lands' (United Nations 1993a: 385). Needless to say, it is often impossible to establish who were the 'original' inhabitants of a region. The term 'traditional' might be interpreted in any number of different ways. Both terms are sometimes defined in ways that prejudge the relationship between people and their environment (see Milton 1996: 201), a relationship which here must be kept open to investigation.

2 The NGOs (non-governmental organizations) involved in producing the document *Caring for the Earth* were the World Conservation Union (IUCN), the United Nations Environment Programme (UNEP) and the Worldwide Fund for Nature (WWF).

3 The *International Treaty between NGOs and Indigenous Peoples* was one of about thirty 'alternative' treaties produced at the Global Forum, the meeting of NGOs held at Rio in 1992 alongside the Earth Summit. The text of the treaties is published in Sutherland (1992).

4 The production of ethnography is itself an area of debate within anthropology (see Crick 1982; Clifford 1986).

5 It is not always clear what anthropologists understand by cultural relativism. It is often taken to mean that cultures can only be properly understood 'in their own terms' (Holy and Stuchlik 1981: 29), and to imply that all cultures are equally valid interpretations of reality and all equally worthy of respect. Cultural constructivism is the view that people 'construct' their view of the world through social interaction (see Milton 1996: chapters 1 and 2, for a discussion of these concepts).

6 Dwyer uses the term 'invisible world', but in view of the fact that some spiritual beings 'may be seen', this term is not entirely appropriate, as he implicitly acknowledges through his considered explication (Dwyer 1996: 163).

8 Aestheticism and environmentalism

David E. Cooper

I

'My parents were ... *intelligents*. And consequently they had the required subtle spiritual make-up. They liked art and beauty ... [with] a special affinity for music.' Shostakovich's (1987: 3) tribute to his parents bears witness to the idea that aesthetic sensibility is an aspect of our 'spiritual make-up'. He does not intend 'spiritual' in a philosophically heavyweight sense here – as meaning that man is a creature of God, say, or possessed of a soul in addition to a mind and body. The point, rather, is that people do not live by bread and beer alone, that for a life to flourish not only must there be satisfaction of everyday, material wants, but fulfilment of capacities to understand, reflect and appreciate. In a low-keyed sense of the term, people seek 'spiritual' fulfilment in exercising such capacities.

It will follow, again in this low-keyed sense, that in aesthetic appreciation of nature a person's 'spiritual' aspect is engaged. (At any rate, it will on a certain understanding of what aesthetic appreciation involves.) And this means that in nature's power to invite such appreciation, we may find a prime reason why our environment should matter to us and be something to 'preserve' out of motives quite different from pragmatic or utilitarian ones, such as conserving resources for future use. Such, at any rate, is the view I want to defend. Those in search of reasons of a non-utilitarian kind to care about the damage we do to nature, about the disappearance of natural landscapes and animal species, do well to cite the aesthetic depredation these entail.

It may sound odd to speak of *defending* such a view. Don't most of us appreciate the majesty of mountains, the serenity of lakes, the grace of flying geese? And isn't that an obvious reason for caring about them – for worrying about yet another ski resort, water-sports lido, or cull of 'pests'? Strangely, however, the idea that environmental care should be founded on aesthetic concerns has received a bad press among many environmental thinkers, especially those of a 'deep green' ecological hue. To the occasional recognition that 'the ultimate historical foundations of nature preservation are aesthetic' (Hargrove 1989: 168), a common response is that it is time to become more enlightened about what the foundations should be than our forefathers were. If aesthetic concerns are conceded to be relevant at all, the concession is typically grudging. It may be better than nothing that the ecologically unenlightened want to preserve certain areas because of their beauty; but better still would be the relegation of aesthetic considerations – as well as more patently pragmatic ones – in favour of 'ecocentric' ones.

Several objections have been raised to the aesthetic grounding of environmental concern – 'aestheticism', as I'll call it[1] – ranging from the charge that aesthetic appreciation is an insufficiently weighty matter to bear the load of an environmental ethic to the claim that appreciation of nature is too different in kind from appreciation of art for us to apply 'aesthetic' to the former. But I begin with the most frequently voiced worry, already indicated in my mention of 'ecocentric' considerations. In the usual (and unfortunate) rhetoric of the debate, the objection is that aesthetic values are 'anthropocentric' or 'humanist', whereas the values environmentalists should be focusing on are those 'intrinsic' to nature. As one author puts it, 'aesthetics . . . cannot form the basis of an adequate environmental philosophy without presupposing that natural processes and their products have no role to play independent of the human evaluation of them in terms of their beauty'. And this, for her, means that 'aestheticism' is a form of 'instrumentalism', according to which 'the natural landscape has as such no intrinsic value – its value lies solely in . . . affording us an aesthetic experience' (Lee 1995: 220). Other writers complain, in similar vein, that an aesthetic approach 'confirm[s] our anthropocentrism by suggesting that nature exists to please as well as to serve us', and that 'there is . . . arrogance in experiencing nature in the categories of art', for it suggests that nature has been 'arranged for the sake of . . . man's aesthetic pleasure' (R. Rees, R. A. Smith and C. M. Smith, quoted in Carlson 1993: 144, 147).

These charges are often backed up by historical considerations purporting to show that aesthetic appreciation of nature – allegedly a pretty recent phenomenon – was derivative from art appreciation, and hence has always been infected by the 'anthropocentric' conceit that nature may be regarded as an artefact. One is reminded, for instance, how in the eighteenth century natural landscapes tended to be admired only to the extent that they obeyed the canons of landscape painting – a tendency illustrated by the use of the 'Claude-glass', a kind of mirror (named after the French painter) for framing the view of the scenery behind one's back. Interestingly, however, the 'fathers' of 'deep' environmentalism do not seem to have shared contemporary fears of regarding nature aesthetically. John Muir applauded Yosemite's resemblance to an 'artificial landscape garden' (Muir 1977: 3), while Aldo Leopold thought 'our ability to perceive quality in nature' must begin, 'as in art, with the pretty', though we soon come to admire further qualities (Leopold 1949: 96). In fact, as Simon Schama's *Landscape and Memory* (1995) demonstrates, the story of appreciation of nature is immensely complicated and forbids glib generalizations about its derivativeness from art appreciation. For example, Dutch landscape painting reflected, as much as it shaped, certain ideals of nature rooted in conceptions of human beings and the symbolic significance of their environments.

What are we to say of the charge that 'aestheticism' represents an 'anthropocentric' denial of 'intrinsic' values in nature? As is apparent in the passages I cited, the central complaint is that 'aestheticism' is a form of 'instrumentalism', in which nature matters only for the delights it affords us, not 'intrinsically' or 'in itself'. Strangely, though, those who press this charge tend to expend most of their energy on a different issue: namely, whether the value of nature is 'intrinsic' in the sense of being 'independent of human evaluation'. 'Strangely', since from the claim that values presuppose evaluation, it hardly follows, just like that, that things have value only as 'instruments' for satisfying human wants. The first claim is a 'formal' one about the concept of value; the latter a 'substantial' claim about what makes something valuable (Eckersley 1992: 55*ff*).

Since mountains, lakes or flying geese can only be described as 'majestic', 'serene' or 'graceful' in virtue of some capacity to affect us, I take it as obvious that nature's aesthetic qualities are 'anthropocentric' in the 'formal' sense of presupposing a relation to human beings. It is important, therefore, for any defence of 'aestheticism' to dissociate this 'formal anthropocentrism' from the 'substantial', 'instrumentalist' kind. Some authors, I suspect, are blind to this ambiguity in the terms 'intrinsic' and 'anthropocentric'; others, while alert to it, think that 'formal anthropocentrism' nevertheless results in the 'substantial' version. And perhaps it is not so very hard to slide from saying that if something's value is not 'intrinsic' and therefore not 'independent', then it must depend on the thing's 'instrumental' contribution to something else.

This slide is greatly abetted by implications which are often, though illegitimately, read into 'anthropocentrism' of the 'formal' kind. It is said to imply, *inter alia*, that values are 'human products', that we 'endow' things with value or 'confer' it on them, and that the things themselves are 'valueless'. Hence the keenness of many environmental philosophers to argue that 'value is not just a human product' and that we should be 'alerted to look for ... natural productions of value' (Rolston III 1986: 124). But such arguments are otiose, for 'formal anthropocentism' does not have the implications imagined. To suppose it does is like imagining that, since things can only be described as loud or soft in virtue of their relation to creatures who are auditorily affected, it is therefore these creatures who 'produce' sounds or 'confer' them on a 'really' soundless world. In the cases of both value and sound, the drawing of such implications confuses quite different senses in which such phenomena 'depend' on us – a conceptual and a causal sense. If we were the 'producers' of value, importing it into an otherwise valueless world, we would indeed be the source of, the causal agency responsible for, the existence of value. And from there it would be tempting to move to the 'instrumentalist' conclusion that values, like other human artefacts, are made in order to further our aims.

But, as the analogy with sounds shows, this is not at all the 'formal anthropocentric' idea: which is, simply, that no sense can be made of things having value in abstraction from actual or potential evaluators. This idea is silent about why we value things: it says nothing about the division of (causal) responsibility between ourselves and the nature of things when things are valued by us.[2] For the wiser proponents of this idea, indeed, reificatory talk of 'values' is dangerous, generating such bad questions as whether these values are 'stuck on' things by us, like so many labels, or are found 'already there', 'in' the things themselves (Heidegger 1980: 190–1). We do better to speak, instead, of the reasons there are for valuing things, leaving open in each particular case what weight is to be attached to the 'intrinsic' qualities of a thing and to its playing a particular role in relation to our wants, aspirations or whatever.

There is one important implication which 'formal anthropocentrism' does have, however, and a thoroughly welcome one. If we are even to make sense of ascribing value to something on the basis of some qualities it has, we must be able to understand why something with those qualities should matter to us, how it might fit into the orbit of our concerns and intelligibly engage with some conception of a good life. This is not a constraint, unfortunately, which is always obeyed by environmentalists who exhort us to value and 'respect' something simply because it is 'natural' or, worse still, simply because it *is*. The *mere* fact that something exists, or exists as a natural phenomenon, has no intelligible bearing on what matters to us and on any conception of the good. Nor, relatedly, could it have any implications for environmental practice and policy. By way of contrast, we readily understand someone who values something for its aesthetic qualities: for although there is disagreement as to how and why majesty, serenity, and grace should matter, no one doubts that they do.

II

I have argued that 'aestheticism', while it treats value in a 'formally anthropocentric' way, is not for that reason a 'substantially anthropocentric', 'instrumentalist' attitude towards nature. But someone might argue that it should, for other reasons, be branded 'instrumentalist'. Moreover, there are other objections to 'aestheticism' I have not yet considered. Some of these can be gathered, together with the 'instrumentalist' charge, into a group, for they all rest on an impoverished and perhaps incoherent view of aesthetic appreciation.

Recall the terminology in which the 'instrumentalist' charge was lev-
elled. For the 'aestheticist', it was urged, nature's value resides 'solely in
affording us an aesthetic experience', in 'triggering off' such experience,
in providing us with 'aesthetic pleasure'. Implicit, here, is an all-too-
familiar model of aesthetic appreciation, according to which something is
admired only because it produces in us certain effects, 'aesthetic experiences', understood as certain 'pleasures'. On this model, it is easy to see
why 'aestheticism' should be classified as 'instrumentalism'. (Indeed, the
model could be labelled 'the instrumental model'.) As money is an instrument for obtaining commodities, aesthetic objects are instruments for
obtaining pleasures. But the same model, I suspect, lurks behind three
other objections to 'aestheticism': they might be dubbed the 'fragility',
'fakery' and 'triviality' objections.

By the 'fragility' objection, I mean the idea that aesthetic appreciation,
given its proneness to variations and shifts in taste, is too insecure a basis
on which to rest environmental protection. One author, for example,
referring to a rugged Lake District landscape, worries that, were there to
be a 'reversion to the pre-industrialization view of such landscapes . . .
they would lose what aesthetic value they now have for us and would no
longer be held valuable' (Lee 1995: 220). Actually, it's hardly obvious
that 'aestheticism' provides a more fragile basis than other suggested
ones, such as 'respect' for humanly uncontaminated nature or ascribing
moral rights to natural objects. For all I know, such respect and ascriptions
will turn out to be more ephemeral than aesthetic judgements on landscapes.

More important, it is easy to exaggerate the ephemerality of the latter
if one operates with the 'instrumental' model of aesthetic appreciation.
For one thing, it is perfectly possible for admiration of something to
survive even if few people derive pleasure from encountering it. And this
is because *pleasure*, in any serious sense of the term, is not the whole,
nor often a part, of what we look for. Someone who finds Goya's 'black'
paintings – or the awesome progress of a tornado – 'pleasing' is perverse.
Shifts in the pleasures people receive may barely affect their aesthetic
judgement. I doubt if we today get the pleasure and amusement from
Aristophanes' plays that the Greeks did: that does not preclude recognizing his merit. Second, this point can be generalized to other
'experiences' which, according to the model, aesthetic objects serve to
provide. The idea that such objects are always and mainly valued for
yielding aesthetic experiences, pleasurable or otherwise, is a bad one. In
appreciating a piece of music, say, for its ingenious structure, the way it
expresses an emotion, or its challenge to a tradition gone stale, I am not
pointing to any 'experiences', let alone pleasures, the music produces in
me. There is a word for someone who looks to music only for the
'experiences' – the lumps in the throat, or whatever – which it can
produce: 'philistine'.

The relevance of this point is that aesthetic appreciation is less 'fragile' than might be imagined. As David Hume (1965: 249*ff*) might have put it, it can survive many differences in the 'humours and manners' of people, for we are often able, sometimes only with effort, to appreciate something's merit despite our inability to share the 'experiences' which our forefathers or the members of a very different culture derived from it. That recognition of such merit may have little or nothing to do with an object causing such 'experiences' is why different cultures, sharing relatively little in the way of taste, can nevertheless admire one another's art. If this point can be carried over to aesthetic appreciation of nature, the message should be cheering to environmentalists. Unlike differences in moral commitments, differences in taste do not demand rejection of all but one's own. My pleasures may be 'pre-industrialization' ones in 'pretty' landscapes; but I do not dismiss people's taste for rugged mountainscapes for, with imagination, I can understand how they find expression, form and meaning in them, even though my own reaction is to feel scared stiff and return to the cultivated valleys below (see Lynch 1996: 156–7 for a similar point).

The 'fakery' objection goes like this: if the only reasons for preserving certain trees, say, were aesthetic ones, then it couldn't matter if the trees were replaced by plastic ones which looked the same and hence offered the same delectation. As one philosopher puts it, if 'we reason from [aesthetic] pleasures', we face the 'appalling implication that natural environments should be replaced by plastic ones'. Even if the notion of the aesthetic is broadened to include more than pleasure, the plastic tree scenario remains a 'risk', since 'trees are only [being] considered instrumentally and accorded no value of their own' (Attfield 1983: 147–8).

This is an odd argument. For one thing, it is not only the look of trees which inspires sensory appreciation, but how they move, smell and feel to the touch. To this it might reasonably be replied that it would be scarcely less 'appalling' to destroy a certain grove of trees and replant with actual imported trees sharing all the sensory qualities of the originals. The irony is that the best way to understand why this, too, might be appalling is by drawing just the analogy with artworks which the 'fakery' objection implicitly denies. For artworks are paradigmatic examples of objects which we are *not*, generally, willing to see substituted by even the most skilful fakes. That we are not is, moreover, a decisive reason for rejecting the 'instrumental' model of aesthetic appreciation: if the value of a painting consisted entirely in its production of pleasures or other experiences in us, it would indeed be hard to see why we should care about it being substituted (See Elliot 1994).

It is obvious that the 'fakery' objection relies on this poor model. It is because aesthetic reasons for caring about trees are allegedly 'instrumental' or 'consequentialist' that they have the plastic-substitute scenario as their 'appalling implication'. The objection dissolves once it is conceded, as it must be given the analogy with artworks, that objects are not appreciated solely, or even primarily, for their production of effects in us that fakes could also produce. It could be objected, here, that the main reasons we do not tolerate fakes in the artworld do not carry over to the natural world – the importance, for example, attached to a painting's originality. This raises the question I turn to in Section III of what is involved in aesthetic appreciation of nature. For the moment, I simply remark that there is no reason to suppose that what is involved does not include consideration of features – something's history, for example – which replicas cannot possess.

Finally, in this group, is the 'triviality' objection. Even if they do not share Baird Callicott's (1993: 151) view that – until the emergence of Leopold's 'land aesthetic' – 'Western appreciation of natural beauty' was 'superficial and narcissistic ... in a word ... trivial', people often say that aesthetic considerations are insufficiently weighty to employ in defence of nature. Isn't aesthetic pleasure icing on the cake of life, not something to be compared in gravity with, in particular, life's moral demands?

Lurking here are a couple of confusions. First, it can be misleading to speak of weighing moral against aesthetic considerations. Moral considerations typically arise when people's interests are at stake, and in some cases these interests are aesthetic ones. It is an aesthetic question whether a certain painting is any good: but the question of whether it is permissible to destroy it, to which its aesthetic quality is relevant, is a moral one. An analogous point could be made about a natural landscape. So the 'triviality' charge should not take the form of elevating moral over aesthetic value, but of arguing that aesthetic interests are unimportant compared to other interests – material needs, say, or personal freedom – when it comes to making decisions. But now, second, we need to distinguish between the question of whether aesthetic interests *are* given importance and the question of whether they *ought* to be.

The answer to the first question is that it depends on which people, when and where, one is talking about. In Stalin's USSR, let's assume, aesthetic concerns were not taken seriously, being strictly subordinated to political and economic ones. In Heian Japan, on the other hand, the aristocracy, at least, attached paramount importance to the aesthetic dimension of life. There have been, in fact, wild variations in the place afforded to aesthetic enjoyment and appreciation in the good life. To the second question, of the place they should be afforded, there can be no quick answer either. Someone who wants to press the case for the centrality of the aesthetic dimension of life must enter the lists with advocates of the importance of other dimensions. To amend an example of Kant's, there can and should be debate over whether a certain building should be erected, given that, while it would be beautiful, it would divert resources from welfare programmes and would anyway be unappreciated by the 'masses'. But there is no *a priori* reason to suppose that the advocate of the aesthetic is incapable of persuading his fellows of the important place of art and beauty in human life.

It will, however, be hard for him to argue a persuasive case if he is lumbered with the 'instrumental' model. Walter Pater thought that life should aim at the accumulation of 'delicious sensations', but few would agree that, if this is all that artworks or natural landscapes provide, they deserve a central place in our concerns. It is just this model, however, which inspires the 'triviality' objection. Once it is stressed, yet again, that aesthetic appreciation typically has little to do with an object's capacity to stimulate 'pleasures', 'art experiences' or 'delicious sensations', the force of the objection is dissipated. Which is not to deny that the case for the importance of the aesthetic, in particular of a natural aesthetic, remains to be made.

III

I have argued that various objections to 'aestheticism' fail, ones which owed any apparent plausibility to the assumption of a poor, 'instrumental' model of appreciation. Once that model is abandoned, the 'instrumentalist' charge against 'aestheticism' must also be dropped. Still, so far I have said little, and only in passing, about the character of aesthetic appreciation of nature, and I have sometimes relied on presumed similarities between this and art appreciation. In this final section, I want to expand on the character of natural appreciation in the course of considering some alleged disanalogies between art and nature.

It would be boring to deny that nature can be aesthetically regarded on the basis, simply, of a stipulative definition of 'aesthetic' as 'pertaining to the arts'. Such a stipulation, incidentally, would be without historical warrant. Baumgarten, Kant and other eighteenth-century writers who first used 'aesthetic' in anything like its present sense happily applied it to judgements of both art and natural beauty. But there are, of course, obvious differences which entail that not all dimensions of art appreciation carry over to nature. Most of these stem from the point that Hegel made, when defending his own restriction of 'aesthetics' to 'the philosophy of the fine arts' – namely, that art is 'born of the spirit' or mind (Hegel 1979: 2). Thus, a tree, unlike a painting, cannot be admired as an outstanding example of a cultural genre.

But some differences cited in order to drive a wedge between art and nature appreciation are illusory. It's been said, for instance, that the former is of particular, discrete objects, like paintings; whereas we are typically 'in' the natural environments that inspire admiration, one not mainly, or perhaps at all, focused on particular objects. But, in the first place, it is not generally true, even with paintings, that surrounding context is irrelevant. Many people feel that something is lost when a fresco is peeled off the wall of the church which is its 'natural' setting and is put in a museum (see Carlson 1993). Second, the alleged difference would anyway be harder to make out in the case of non-plastic artforms, such as music. The symphony one hears 'surrounds' one: it is not 'over there', a discrete object standing free of other things in the perceptual field.

Before I consider other alleged differences between art and nature appreciation, I want to go on the offensive and offer two (very different) reasons for thinking that any sharp disjunction is implausible. The first is that there are several kinds of case where the attempt to bisect what is appreciated into two components – art and nature – is forced, and perhaps absurd. Take, for example, Japanese gardens, to which few would deny the status of art. As Watsuji Tetsuro pointed out, the Japanese gardener succeeds 'only by making the artificial follow the natural [a]nd by the nursing of the natural by the artificial' (1961: 191). The appeal of a tea-garden may partly owe to the moss whose natural undulation is heightened by skilful placing of stones. It would be impossibly strained, surely, to insist that the viewer here enjoys two items which he discriminates – the natural stuff (the moss) and the gardener's artistry. (If he did, the gardener has failed: see Takashi 1981.) Another kind of example is furnished by certain buildings. Let me here quote Heidegger's perceptive remarks on a Greek temple, which surely imply that bisection of the total experience of the temple and its setting into an experience of art *and* another one of nature would be incoherent:

> Standing there, the building rests on the rocky ground. This resting of the work draws up out of the rock the mystery of that rock's clumsy yet spontaneous support. Standing there, the building holds its ground against the storm raging above it and so first makes the storm itself manifest in its violence. The lustre and gleam of the stone ... first brings to light the light of the day, the breadth of the sky, the darkness of the night. The temple's firm towering makes visible the invisible space of air. The steadfastness of the work contrasts with the surge of the surf, and its own repose brings out the raging of the sea.
>
> (Heidegger 1971: 42)

My second reason for denying a disjunction appeals to the thought, elaborated by Kant (1952: 42), that the distinctive mark of aesthetic appreciation or 'the judgement of taste' is that it is 'independent of all interest'. Such appreciation is 'disinterested' in that, unlike appreciation of a hot bath after a game of rugby or of a jury's just verdict, it is not due to the satisfaction of antecedent desires or 'interests' – physical, moral or whatever. This thought needs both protecting against misunderstanding and considerable fleshing out. Thus, it should not be understood as excluding moral concerns from a role in art appreciation: but we need to distinguish, here, between admiring a painting for its moral message and doing so for the skill with which that message is communicated. The former is a moral response, the latter an aesthetic one. And the aesthetic response is not available to someone incapable of setting aside or standing back from, however briefly, their moral commitments and emotions, incapable of 'disinterestedness'.

A promising way to flesh the thought out is by introducing the idea of the 'alternative worlds' which great works of art 'set up' and invite exploration of. Consider, say, the 'alternative world' one enters when reading a great novel. This is an *alternative* world since, as Alan Goldman (1995: 151) puts it, we 'lose our ordinary, practically oriented selves in the world of the work'. (Which is not to deny that we can return from it with new perspectives on our actual world.) And it is an alternative *world* because of the rich and diverse dimensions provided for our exploration – formal, emotional, narrative, symbolic, and so on. The approach is promising since it at least gestures towards an explanation of why art matters to us. For surely there is a 'value of entering other worlds for its own sake', of employing our 'sensory, cognitive [and] affective' faculties in relative freedom from the pressing demands of life, but also a value of such an entry 'for its refreshing effects on [those] faculties' (ibid.: 155), and for what we may bear back with us from an 'alternative world' to the actual one.

What has this to do with appreciation of nature? Kant rightly insisted that nature, possibly more than art, affords 'disinterested' contemplation and admiration, and was entitled therefore to bring both art and nature appreciation under the heading of the aesthetic. Put in terms of the 'fleshed out' version, natural environments, too, can 'set up' or constitute 'alternative worlds' to explore with the full range of one's faculties. Sometimes, a person who 'goes into nature' is, somewhat literally, entering a different world from the city which is the milieu of his or her everyday, practically engaged existence. But even when a forest, say, *is* the milieu of a person's 'practical orientation' – as with a charcoal-burner – it can become an 'alternative world'. Just as someone is very differently related to the sounds of an orchestra in the capacities of sound-engineer and music-lover, so someone is differently related to the forest which is the source of his livelihood and to the same forest when, on a Sunday, tarrying on its paths, smelling its resin, and watching the hawks soar above it. In my terminology, a different 'world' is being explored from the workaday one of practical involvement.

We are now in a better position to see what people must argue who want to drive a wedge between art and nature appreciation. They must reject my claim that a natural environment can be an 'alternative world', inviting exploration by all our faculties. Presumably, though, they will concede that nature, as much as art, furnishes forms and materials to engage our interest in structure, our sense of colour, balance and so on. Indeed, this seems the one thing that is conceded by those who, employing the 'instrumental' model, refer to the 'aesthetic pleasures' that landscapes 'produce' in us. Let's simply note the concession and pass on, pausing only to deplore the reduction of the complex, active and imaginative exploration of formal relations between shapes and colours, be they in the sky or on a Mondrian canvas, to 'receiving aesthetic pleasure'.

But what of the many other dimensions of which a great artwork invites exploration and contemplation, and in virtue of which the metaphor of an 'alternative world' is apposite? For example, the dimensions of expression, meaningfulness and relatedness to the culture of the actual world. Surely in the experience of a lake, there is nothing like identifying the feeling a painting expresses, a belief it communicates, or its place within a certain tradition!

I suggest that there are such similarities, and only bad claims about art or about our experience of natural places precludes recognition of them. Let me begin with *expression*. According to a popular view, a painting expresses a feeling because it is the one the artist felt and wanted to communicate. If this were true, it would indeed make no sense to speak of something natural expressing a feeling. But it is not true. A piece of music may be sad, expressive of sadness, despite the fact that the composer felt on top of the world when writing it. I don't pretend to have an adequate account of what it is, exactly, for music to express sadness. The view that it is a matter of arousing sadness in listeners is certainly too crude: I can hear sadness in Sibelius's 7th Symphony without myself becoming sad. But the relevant point is that, whatever the adequate account might be, there is no reason to deny that natural phenomena can be expressive. That they do not exhibit a maker's feelings is beside the point, for nor need paintings which are expressive of feelings. In fact we speak quite naturally of the sadness of a wintry moon, the exuberance of a tumbling stream, and the like. Moreover, just as a good critic may discern in a piece of music feelings which listeners have overlooked, so the sensitive explorer of nature may persuade us to revise our responses. Maybe it's not sadness, but calm resignation before the process of life – what the Japanese call *wabi* – which the bowing reeds express.

These remarks on expression apply as well to the notion of meaning. One does not have to embrace today's fashionable 'role of the reader' or 'death of the author' thesis – with its insistence that meaning is freely 'projected' on to 'texts' by their audiences – to accept that meaning in art is not simply a function of the artist's intention. (If it were, there could be no ascription of meaning to natural phenomena.) Clearly works can have a significance – convey a 'message', perhaps – that even the artists themselves may concede was not part of their intention. This may be due – though it need not be – to their employing words, sounds or shapes which, willy-nilly, are bound to have a certain 'conventional' significance for an audience.

Once the tie between meaning and intention is loosened, there is no compulsive reason to deny, in general, that natural phenomena may have meaning. Indeed, there is a very long tradition, in both West and East, of discerning metaphors, symbols and so meanings in nature. Think of the medieval idea of nature as 'the Book of God' to be 'read' by those who know its 'language'. And who could deny, to use a clichéd example, that in nature's 'spring awakening' men and women have always found a metaphor for human awakening, for birth, and for making fresh starts in life? I don't, though, want to create the impression which it is easy to get from Lévi-Strauss's well-known remark that animals (and plants and berries) are 'not only good to eat, but good to think with'. That has too 'instrumental' a ring, as if one just 'used' nature, in the way one does a mnemonic or an abacus, to help one think and reflect. This is like supposing that all readers of a great novel must be like the sociologist who scours it for what it may signify about the author's society. That the discernment of meaning is an important aspect of aesthetic appreciation does not imply that artworks or nature are to be approached with the motive of finding metaphors or symbols. On the contrary, without a 'disinterestedness' that excludes such a motive, one cannot be open to much of the significance that is there to discern.

That environments are already imbued with meanings that generations of people have 'read' them as having is one reason why, in the appreciation of nature, there is at least an analogue to exploring the 'alternative world' of an artwork for its relation to the actual world – to, for instance, recognizing that a genre of painting typifies its age. More generally, there are few natural environments – not even all those without any mark of human intervention – in which there are not to be found what Schama (1995: 14) calls 'the veins of memory'. For nearly all environments belong to history in the sense of being connected with ancient and changing visions of the natural world and its relationship to human beings. Only someone either ignorant or of stunted spirit could view the natural setting of Heidegger's Greek temple, respond with 'Very nice, too!', and remain oblivious of the relation of sea, hills, rocks to the Greeks themselves, a relation which helped to give to these 'men their outlook on themselves' (Heidegger 1971: 42).

I have argued, then, that no wedge should be driven between the appreciation of art and of nature. And I hope I have said enough about the character of such appreciation to defend the idea that aesthetic appreciation belongs, as Shostakovich put it, to people's 'spiritual make-up'. It *may* be that spirit is involved in a more 'heavyweight' sense as well: maybe Daisetz Suzuki (1973: 363) was right to say that 'the appreciation of beauty is at bottom religious ... without being religious one cannot detect ... what is genuinely beautiful', not all of it anyway. But there is no need to insist on that in order to recognize that people without aesthetic appreciation of nature are stunted in spirit. I find it hard to think of a more important reason for cultivating concern for our natural environments and for what is currently befalling them.

NOTES

1 'Aestheticism' is not, perhaps, an ideal term, since it also used, pejoratively, to refer to an 'art for art's sake' effetism. Although I am not advocating that, I do want to retain one important connotation of 'aestheticism' so understood – the opposition to 'instrumentalist' views of art.

2 A curious byproduct of the failure to heed the distinction between conceptual and causal dependence is the 'last man' argument, which purports to show that values are 'in' nature, independently of human valuation (see, for example, Routley and Routley 1980). Surely, the argument goes, it would be wrong of the last man on earth to devastate it just before dying – something, it is alleged, the 'anthropocentrist' cannot explain since, for him, it can't matter what a human-free, 'valueless' earth is like. This picture of the world suddenly becoming valueless is as quaint as that of it suddenly becoming soundless when the 'last hearer' dies. It would be mad to hold that human beings are responsible for value and sound as they are for, say, vegetable prices, which will indeed disappear when we do. The conceptual dependence of value and sound on valuers and hearers is, however, perfectly compatible with speaking of them in a humanless world. Nor is there any reason why, despite this dependence, the last man should be indifferent to the earth's fate. If he really does value this river, say, how could he be? Not wanting it to dry up, now or later, is part of really caring about it.

9 The Romantics' view of nature

Greg Garrard

I could not bear to see the tearing plough
Root up and steal the Forest from the poor,
But leave to freedom all she loves, untamed,
The Forest walk enjoyed and loved by all!

(Clare 1986: 165)

Friday 23rd April 1802. . . . After we had lingered long looking into the
vales – Ambleside vale with the copses the village under the hill &
the green fields – Rydale with a lake all alive & glittering yet but little
stirred by breezes, & our own dear Grasmere first making a little round
lake of natures own with never a house never a green field – but the
copses and the bare hills, enclosing it & the river flowing out of it.
Above rose Coniston Fells in their own shape and colour – not Man's
hills but all for themselves the sky & the clouds & a few wild creatures.

(D. Wordsworth 1991: 90)

In literary terms, these are two voices from the margins – the first a
labourer called John Clare who made some of the most important contributions to Romantic proto-ecological writing, and the second Dorothy
Wordsworth recording in a journal a walk in the Lake District with her
brother William and her friend Samuel Taylor Coleridge. Both are relatively minor figures in the vast, disconnected phenomenon called
'Romanticism', which I will be treating solely in its British and German
manifestations in the period 1789–1837. There has been considerable
argument among literary scholars over whether there are enough – or
even any – unifying features in what we call 'Romanticism' to justify the
name. Certainly, the British 'Romantics' never called themselves by that
name – when William Wordsworth, for example, uses the term (see p.
119 below) he means it in the pejorative adjectival sense of 'unreal' that
we still use today. None the less, it seems to me that – to borrow
Wittgenstein's metaphor – there are overlapping 'family resemblances'
between various 'Romantics' right across Europe that allow us the cautious use of the term. One such feature is evident from these quotations:
an unprecedented insistence on a kind of intrinsic value in nature, a

worth not reducible to an instrumental calculus of resource base or agricultural potential. At the same time, we can begin to get an inkling already of the extraordinary *variousness*, and at times contradictoriness, of the Romantics' views of nature that I propose to discuss; Clare wants Epping Forest saved for the sake of the reciprocal love between place and people that he claims exists, while the Fells are of value for Dorothy Wordsworth precisely insofar as they are 'not Man's hills' at all.

Certainly an apprehension of the epochal significance of Romantic writing in terms of a new view of nature is not novel; Raymond Williams identifies it in *The Country and the City* as an 'active sympathy' to be distinguished from the mixture in the earlier part of the eighteenth century of 'agrarian confidence' and 'feelings of loss and melancholy and regret':

> Now, with Wordsworth, an alternative principle was to be powerfully asserted: a confidence in nature, in its own workings, which at least in the beginning was also a broader, a more humane confidence in men.
>
> (Williams 1993: 127)

William Wordsworth himself claimed, in the Preface to the 1802 edition of *Lyrical Ballads*, that the poet – by which he meant the proper poet, or himself – 'considers man and nature as essentially adapted to each other', depicting incidents of rural life especially because 'in that condition the passions of men are incorporated with the beautiful and permanent forms of nature' (Wordsworth and Coleridge 1991: 259, 245). This, he supposed, distinguished him from many of the poets of the previous age, who were often concerned more with urban mores than with the relationship of humans with nature. The vital terms here are 'adapted' and 'incorporated', because they distinguish Romantic sympathy – which is essentially a spiritual reciprocity – from the scientific or ecological sense in which they would be used today.

Romantics wrote at a critical time in European environmental history. The Germans were mainly bourgeois intellectuals[1] whose class was stifled by a political geography of fragmented, feudal states, yet simultaneously fired by the French Revolution and the Napoleonic occupation. Their interest in nature was usually an adjunct to strongly aesthetic preoccupations, perhaps because industrialization was neither as widespread nor as advanced as in Britain. In these islands, scientific discoveries had shaken the strongly anthropocentric outlook of the early modern period, in a process we now call the 'decentring of the human subject' – anatomical and physiological likeness appeared to suggest, long before Charles Darwin,[2] a close biological relationship between 'men' and 'brutes', which cast doubt upon the special status apparently given to Adam's progeny in the Bible. Eighteenth-century naturalists had increasingly sought to distinguish their work – which tried to give accounts of nature as in itself it truly was – from the traditions and superstitions of rural people which constantly related the phenomena of the non-human world to the human or divine phenomena they were supposed to symbolize or prefigure, and while this led to greater understanding of life on earth (and ultimately to the birth of ecology) it also inevitably tended to disenchant and reify it. Clearly industrial exploitation of nature was enabled and enhanced by this historical transformation, and also tended reciprocally to reinforce its materialistic outlook. Keith Thomas has given an excellent account of this process in *Man and the Natural World*, which concludes at the dawn of the Romantic Age in 1800 with the claim that, in the previous centuries,

> there had gradually emerged attitudes to the natural world which were essentially incompatible with the direction in which English society was moving. The growth of towns had led to a new longing for the country-side. The progress of cultivation had fostered a taste for weeds, mountains and unsubdued nature. The new-found security from wild animals had generated an increasing concern to protect birds and pre-serve wild creatures in their natural state.

> (Thomas 1984: 301)

So we find ourselves, in certain respects at least, with the Romantics as our contemporaries. As environmentalists, we hold beliefs at variance – often to a drastic degree – with those which constitute the ruling ideology. Moreover, when we propose contemporary nature writing as an element in a programme of environmental education, we appear to be subject to the same difficulty of interpretation: ought Romanticism or modern 'ecological' writing be seen as genuine challenges, or as varieties of 'false consciousness', which function only to relieve the guilty con-science of a destructive society while masking its real operations? There are no neat, universally applicable answers, and ambivalence is rife; Wordsworth may be credited with having inspired the creation of National Parks, but he also helped to promulgate the aesthetic which has filled Cumbria with tourists and exposed less picturesque (but perhaps more valuable) environments such as wetlands and meadows to almost uncon-trolled exploitation. So we cannot assume that writing that celebrates nature will be any panacea in the environmentalist struggle. None the less, Romantic writing remains crucial for an understanding of the historical genesis of environmental concern, and also stands as a spiritual resource, a wellspring of love for, sympathy with and confidence in the natural world, without which no ethical or political enterprise can hope to succeed.

BLAKE'S TYGER, BURNS'S MOUSE

One major element of the ethical and ontological shift in the eighteenth century was the transformation of Britain from a byword for cruelty towards animals into a renowned haven of humane sentiment (or womanish pity, in Spinoza's terms, to take the opposing view). Although by no means universal or effective, this change can be seen as an essential part of the historical movement towards a more inclusive ethic. By this I mean that the 'widening circle' of moral considerability appears to have become less and less parochial over time, gradually bringing in 'outgroups' such as women, slaves, disabled humans, animals and eventually perhaps plants and ecosystems.[3] The Romantic period saw some of the most significant expressions of the tentative steps beyond 'human chauvinism', although it also tended to highlight some of the ambivalence of the move that remains today.

Burns's poem 'To a Mouse (on Turning her up in her Nest with the Plough, November, 1785)' may be taken – indeed has been taken – as an exemplary effusion of sympathy for the 'wee, sleekit, cowerin, timorous beastie' (Burns 1968: 127–8). The poet is disturbed by the terror of the mouse, which follows from the way in which human dominance 'has broken Nature's social union', although here the dominion itself remains unchallenged. Still, it would have to be qualified by the creature's stated kinship with his 'poor earth-born companion / An' fellow-mortal' the poet, and by the knowledge that the latter has of the dire consequences that will follow for the mouse with its 'wee bit housie' wrecked as winter closes in. Yet the poem does not conclude on a note of sympathy, but rather contrasts the real suffering thus imposed upon the animal with the greater anguish of a human, for whom it may be vastly compounded by temporality: human griefs and fears achieve a qualitative differentiation through the apprehension of past and future.

If Burns's poem is effective for its qualified compassion, Blake's great poem from *Songs of Experience* adopts a very different approach. Far from asserting a kinship, 'The Tyger' takes the beast as a figure of challenging sublimity; the sense of beauty and horror attaching to the 'Tyger! Tyger! burning bright / In the forests of the night' (Wright 1987: 74) reveals it as something quite other than human. We might take this as an example of the minority view that respect for nature demands an acceptance of its inevitable alterity, not our interrelationship with it. However, the poem uses the animal strictly as an opportunity to catechize the Creator: 'What immortal hand or eye / Could frame thy fearful symmetry?' In particular, we can isolate the question that shows how anthropomorphic the poem remains:

When the stars threw down their spears,
And water'd heaven with their tears,
Did he smile his work to see?
Did he who made the Lamb make thee?

(ibid.)

Blake was not the last to confuse cruelty with predation; under an essentially Manichean view of earthly human and bestial existence, the eating habits of the tiger are taken for an issue of theodicy rather than nutrition. Both these poems from the early Romantic period show how, although attitudes were shifting and under interrogation, ambivalence regarding human nature and the cosmos still predominated over consideration of the natural world for its own sake.

DOROTHY AND WILLIAM: HOME AT GRASMERE

In the Wordsworths we have two of the most important figures in any consideration of the Romantic proto-ecological view of nature. We also have some of the most problematic elements of that view. On one hand, William Wordsworth has exerted a profound influence upon the development of English-language nature poetry and the feeling for nature generally. On the other, it cannot be denied that he tended to see nature as a vital but subordinate element in an essential relationship of humanity and divinity, particularly in his later years. Together, however, Dorothy and William[4] reconcile some of these tensions; as William writes, 'she gave me eyes, she gave me ears' (Wordsworth 1987: 62) and with them he saw and heard nature's life in a uniquely significant way.

Lyrical Ballads, co-authored with Coleridge, contains many of William's most affecting and important poems on humanity and nature. In particular, 'Lines written in early spring' has caught the attention of more than one ecocritic, because in it the poet counterposes his faith in nature and the disenchanted mode of perception which is antithetical to it. Moreover, his 'active sympathy' itself provokes despair over the denatured situation of humanity:

> To her fair works did nature link
> The human soul that through me ran;
> And much it griev'd my heart to think
> What man has made of man.

> (Wordsworth and Coleridge 1991: 69)

Karl Kroeber comments on the last line of the stanza:

> In the context of this experience of the natural world . . . it seems fair
> to say that the line suggests not only how we fight with, degrade, and
> destroy one another but also how such self-injuries may be rooted in
> a miseducation of ourselves away from linkages through pleasure with
> beneficent natural processes.

> (Kroeber 1994: 45)

The following three stanzas see the poet wrestling with his ambivalence; on the one hand he claims that 'every flower / Enjoys the air it breathes' and that every movement of the birds 'seem'd a thrill of pleasure', but on the other he admits that he cannot be sure. The thoughts of the birds cannot be measured, and the notion of vegetable pleasure – in the primrose-tufts, the periwinkle and the 'budding twigs' – seems to admit its own implausibility through the poet's insistence that he has no choice but to entertain the idea: 'I *must* think, do all I can, / That there was pleasure there' (ibid.; my emphasis).

If Kroeber emphasizes the poet's resistance to his acculturated alienation from a disenchanted nature, another ecocritic, Robert Harrison, claims that William's poem is a 'testament of civic irony', highlighting even as it laments a necessary ontological distance:

> The poet knows that he too is a man and that, however persuasive his feeling of appurtenance to nature, he is condemned to the legacy of what man has made of man. The one who grieves is already conditioned in advance by the man-made world. If nature originally created man, man in turn takes over the creative process and makes of himself something unearthly.
>
> (Harrison 1993: 158–9)

This is more than just a critic's dispute – it goes to the heart of modern environmental thought. Is our alienation from nature merely a local, historical phenomenon, or is it really necessary, an ontological fact? In 'The world is too much with us', William follows the German aesthetician F. W. Schiller in historicizing human alienation from nature: while 'we' are 'out of tune' with nature, the 'Pagan suckled in a creed outworn' lives in an animate universe where he might 'hear old Triton blow his wreathed horn.' On the other hand, it is arguable – as for instance in Harrison's book *Forests* – that the apparent Romantic *invention* of alienation was really only the first *formulation* of an essential estrangement common to all humans as language users. I have suggested elsewhere that this issue is strictly undecidable,[5] and indeed here William leaves the issue unresolved, but sincerely posed, with a question mark at the end of the last stanza.

It is worth noting here that, despite his use of such terms as 'creed', 'faith' and 'soul', William's spiritual kinship remains without transcendental connotation at this stage in his life, and it is even perhaps pantheistically inclined. However, in *The Prelude*, William evinces a more typical view, which makes claims for humankind's transcendental desire. The 1805 version retains much of the language of the earlier poetry, with frequent references to the 'sanctity of nature', but also asserts with equal conviction that 'our destiny, our nature, and our home, / Is with infinitude – and only there' (Wordsworth 1979: 216). By 1830, when major revisions were done, William was an orthodox Anglican, and nature – although still a thoroughly beneficent influence – was completely subordinated to a transcendent divinity.

Yet the Wordsworths were not only preoccupied with abstract or religious questions. In the period around the turn of the century when most of William's great poetry and Dorothy's journals were written, their main concern was to make a home for themselves together to make up for a childhood spent apart, and to do so in the best place imaginable – their birthplace, England's Lake District. In Grasmere Vale the Words-worths found spiritual peace,

> ... the sense
> Of majesty and beauty and repose,
> A blended holiness of earth and sky ...

> (Wordsworth 1977: 46)

They also found a thriving community, as yet unharmed by the traumatic changes occurring in other parts of Britain. Both poetry and prose are full of paeans to the 'mountain liberty' of their fellow dwellers among the lakes; in *The Prelude* (1805) the heart of every inhabitant is nature's 'dearest fellow-labourer'. There William sees:

> Man free, man working for himself, with choice
> Of time, and place, and object; by his wants,
> His comforts, native occupations, cares,
> Conducted on to individual ends
> Or social ...

> (Wordsworth 1979: 274)

Dorothy's journal records many occasions which demonstrate their sense of belonging to this community, such as their purchase of a good deal of gingerbread from some poor neighbours: 'I could not find in my heart to tell them we were going to make gingerbread ourselves' (Wordsworth 1991: 137).

At the same time, Dorothy's journal is full of episodes of hardship, illness and grief – both their own and others' – and William is similarly careful to avoid what he calls (ironically, from our point of view) 'romantic hope', or evasion of harsh realities. After relating one of many stories of rural trial and fortitude, William asks the key question of pastoral poetry:

> Is there not
> An art, a music, and a stream of words
> That shall be life, the acknowledged voice of life?
> Shall speak of what is done among the fields,
> Done truly there, or felt, *of solid good*
> *And real evil*, yet be sweet withal,
> More grateful, more harmonious than the breath,
> The idle breath of sweetest pipe attuned
> To pastoral fancies?

> (Wordsworth 1977: 76; my emphasis)

Alongside this feeling for the human inhabitants is a protective attitude towards the non-human life of the place. Dorothy writes:

I found a strawberry blossom in a rock, the little slender flower had more courage than the green leaves, for *they* were but half expanded & half grown, but the blossom was spread full out. I uprooted it rashly, & I felt as if I had been committing an outrage, so I planted it again – it will have but a stormy life of it, but let it live if it can.

(Wordsworth 1991: 61)

On a larger scale, William is dismayed upon returning to Grasmere after a nine-month absence to find that the fir trees surrounding the church have been cut down, and writes:

... unfeeling Heart
Had He who could endure that they should fall,
Who spared not them nor spar'd that sycamore high,
The universal glory of the Vale.

(Wordsworth 1986: 42)

This concern with real environmental issues seems to become more evident, even as William's philosophy becomes more conservative and anthropocentric. His most popular work in the nineteenth century was not a work of poetry, but a series of editions of his *Guide to the Lakes*, which combined in its final fifth edition: guided walks, discourse upon human and natural inhabitants, opinions regarding the future conservation and development of the area, a sonnet series on the River Duddon, botanical tables supplied by Thomas Gough of Kendal, geological observations by Reverend Professor Sedgwick and Dorothy's descriptions of walks to Ullswater and up Scafell Pike.[6] Perhaps paradoxically, William's *Guide* was likely to promote what he himself called the 'blight' of mass tourism (although, to be fair, he tried hard to promote environmental awareness in its readers). When a company was formed in 1844 to build a branch line of the Lancaster and Carlisle Railway to Bowness near Lake Windermere, William embarked upon a vigorous campaign against it. One major objection rested upon a theory of aesthetic education: the taste for sublime scenery must be cultivated and gradually acquired by those (city-dwelling workers) not born to it, so the visitors would actually be better off getting to know their own locale, or at least the pleasant countryside surrounding it. Perhaps, after suitable training, such people might be allowed to visit the Lake District. The other objection is stronger, less tainted by class politics, and concerns the familiar problem of dramatic diminishment of aesthetic returns to tourists arriving *en masse*.[7]

What can ... be more absurd, than that either rich or poor should be spared the trouble of travelling by the high roads ... if the unavoidable consequence must be a great disturbance of the retirement, and in many places a destruction of the beauty of the country, which the parties are come in search of? Would not this be pretty much like the child's cutting up his drum to learn where the sound came from?

(Wordsworth 1974: 346)

Once again, we encounter ambiguity: is William urging us to 'love nature, but Not In My Back Yard', or is he a prophet of destructive mass tourism? Perhaps the options are not easily disentangled. In any case, the Wordsworths must remain critical for any understanding of the development of environmental consciousness in Britain, and for an exemplary and complex articulation of it at the outset.

THE RELIGION OF THE FIELDS AND THE THRALDOM OF EARTH

Friedrich von Hardenberg – a.k.a. 'Novalis' – and John Clare could hardly be more different either as people or as poets. The former was an aristocrat who studied philosophy in the period of major post-Kantian activity, and the latter a country labourer raised in poverty and largely self-educated. Through these writers we can perhaps see the differences *in extremis* between two important German and British tendencies in Romanticism.

If William Wordsworth is the poet of 'sublime' nature – of Lakeland mountains and streams – John Clare records the minutiae of the wild nature that survives in the interstices of agriculture. Indeed, far more than any other Romantic writer, his theme is *survival*, and this gives him an unrivalled importance, especially when his unique social position is taken into account: 'His position is very rare: close enough to the soil to experience its cruel intractability and joyous tang, he is sufficiently distanced to see his environment as a whole' (Brownlow 1983: 4). Between the dual insecurities of professional writing and agricultural labouring, in a period of radical changes in both publishing conditions and agro-economics, Clare suffered no lack of personal identification with the threatened marginalia of rural life.

There are many examples available of Clare's sympathy, most often expressed towards threatened trees, vulnerable birds' nests and wild flowers among the arable crops. Non-human subjects are sometimes even compared favourably with their human counterparts, as in his poem 'To a Fallen Elm':

Friend not inanimate – tho stocks and stones
There are and many cloathed in flesh and bones
Thou owned a language by which hearts are stirred
Deeper than by the atribute of words

(Clare 1986: 85)

More extraordinarily still, Clare on occasion gives nature a voice of its own; 'The Lament of Swordy Well' is not written on behalf of a place, but narrated *by* it:

> Though I'm no man yet any wrong
> Some sort of right may seek
> And I am glad if een a song
> Gives me room to speak

(ibid: 94)

The old stone quarry complains that the tyrant 'gain' and its servant parliamentary enclosure have taken away its freedom and independence, and placed it like a pauper on to the mercy of the parish. Much else is lost in this change besides: the wild flowers that flourished there are gone, the wild and domestic animals can no longer graze and 'the gipseys camp [that] was not affraid' before boundaries were imposed has moved on. Swordy Well speaks for itself: 'Ive scarce a nook to call my own / For things that creep or flye.' It mourns the hypocrisy and greed of those who loved and would now exploit it, and recognizes that only the protection of wealth can suffice:

> ... save his Lordships woods that past
> The day of danger dwell
> Of all the fields I am the last
> That my own face can tell

(ibid.: 99)

The danger is that the place will be reduced to a label that signifies nothing: 'My name will quickly be the whole / Thats left of swordy well'. The refuse tip at Swaddy Well, Northamptonshire, is the realization of that fear.

Once again, however, we must exercise some caution. When Clare protests against enclosure he contrasts two modes of 'freedom': the freedom of 'gain' to do as it pleases, and the freedom of wild places like 'The Mores' (moors) that 'never felt the rage of blundering plough'. But he does not just mean freedom from interference; it is also a topographical point:

> In uncheckt shadows of green brown and grey
> Unbounded freedom ruled the wandering scene
> Nor fence of ownership crept in between
> To hide the prospect of the following eye
> Its only bondage was the circling sky

(ibid.: 90)

It is not only this place but its unbroken continuity with the open fields of the pre-enclosure agricultural system that Clare wants to defend. Yet, in another historical irony, it is not that landscape but the hedged patchwork that Clare saw as oppression which is valued and defended as 'timeless' and 'traditional' in England today. The hedges and copses of the landscape of modern nostalgia to Clare make up 'little parcels little minds to please', and trespass on the sacred freedom of the land:

Each little tyrant with his little sign
Shows where man claims earth glows no more divine
On paths to freedom and to childhood dear
A board sticks up to notice 'no road here'
And on the tree with ivy overhung
The hated sign by vulgar taste is hung
As tho the very birds should learn to know
When they go there they must no further go

(ibid.: 92)

The contention that enclosure impinges upon a natural sanctity may be confused – the open field system was less efficient but not noticeably more respectful of nature – but it does highlight an important aspect of Clare's work. For him, the realm of the spirit is neither in a church nor in individual communion with God; he is not convinced that, as Swordy Well says in its 'Lament', 'those who go to church / Are eer the better saints'. Moreover, he finds collective worship uncongenial: 'I got a bad name among the weekly church goers forsaking the churchgoing bell & seeking the religion of the fields tho I did it for no dislike to church for I felt uncomfortable very often' (ibid.: 66–7).

Novalis, on the other hand, clearly took seriously his teacher Fichte's claim that reality – the 'non-I' – is posited only for the sake of the ego or spirit, which is the only certainty and truth. Nature for many German Romantic poets was valued as a foil to consciousness, or as a scene of transcendence. Novalis sought to overcome the dualism of subject and object, or spirit and reality, but not through a conception of material relatedness. Rather, he felt that mystical perception especially available to poets could lead to a mystical oneness of Self and universe. Yet the lack of empirical observation in poetic perception ensured that, like the speculative Romantic science of *Naturphilosophie*, his 'nature' seems an extension or function of consciousness rather than vice versa.

In Novalis's famous *Hymns to the Night* (*Hymnen an die Nacht*), he appears to approve the joyful light, and the living creatures symbolically associated with it:

As a king
It summons each power
Of terrestrial nature
To numberless changes,
And alone doth its presence
Reveal the full splendour
Of earth.

(Flores 1960: 56)

Yet love, and 'joys which are the promise of heaven', belong to the night and longed-for death. The greater part of these verses are, as their title suggests, devoted to the praise of the holy night, and organic life viewed in accordance with the dominant Christian tradition as a temporary, ultimately derivative mode of existence:

> Must ever the morning return?
> Endeth never the thraldom of earth?

(ibid.: 59)

Novalis contrasts present life with a prelapsarian Utopia when:

> Trees and brooks,
> Blossoms and beasts
> Had human sense.

(ibid.: 65)

The contrast with Clare could not be more complete. Where the latter sees a precious, wonderful, threatened natural world, Novalis's religiously inspired vision is of an alien slave awaiting redemption:

> Lonely and lifeless
> Stood Nature
> Robbed of her soul
> By strict number
> And iron chains.

(ibid.: 67)

Not surprisingly, such a place is not too amenable to poetic vision:

> Doth not all that inspires us
> Bear the color of night?

(ibid.: 62)

I think it unlikely that a Christian world view such as this has made much of a contribution to environmental destruction, and in any case it would be unwise to follow some of the more crude Marxist and feminist critics in assessing writers on the basis of some predetermined standard of political commitment. None the less, we may legitimately ask whither our eyes should be turned today: towards the frail hues of actual flowers that have been and may pass away, or beyond, from this 'fallen' world towards the transcendent mystery of the 'color of night'?

FIVE NIGHTINGALES

I have never heard a nightingale sing in Britain. The most beautiful place I have ever been was a river gorge in South-Central Anatolia where you could not see the birds, but could hear little else besides their 'delicious notes', their massed 'tumultuous harmony and fierce' (Wright 1987: 179; Wordsworth 1987: 148). Throughout their European range, nightingales are threatened by hunters and by habitat destruction, giving the lie to Keats's lines:

Thou wast not born for death, immortal Bird!
No hungry generations tread thee down.

(Wright 1987: 278)

Romantic poetry had a particular affinity for nightingales, explicable partly by their fantastic and varied songs. These seem spontaneous effusions, making them ideal analogues for Romantic models of artistic creativity. Coleridge compares the burst of song that saturates the wood when the moon emerges at night to the aeolian wind-harp, another favoured symbol of inspired creation:

... these wakeful birds
Have all burst forth in choral minstrelsy,
As if some sudden gale had swept at once
A hundred airy harps!

(ibid.: 180)

Wordsworth, Clare, Coleridge, John Keats and the German Romantic Joseph von Eichendorff all wrote nightingale poems, two of them among the greatest Romantic writing. Wordsworth's is not one of them; his poem contrasts the nightingale's vigorous song 'in mockery and despite / Of shades and dews, and silent night; / And steady bliss' (Wordsworth 1987: 148) with that of the stock-dove which is a gentle, reassuring cooing, 'slow to begin and never ending'. The song of the dove is the one he favours – 'the song for me' – which seems to me rather ostentatiously 'mature'.

Coleridge contributed his beautiful poem 'The Nightingale' to the *Lyrical Ballads* collection. It is exemplary in its ambivalence: apparently opposed to anthropomorphism, yet also striving towards a joy and confidence in nature. Thus, empirical knowledge can enhance our appreciation of nature which might otherwise be limited by stereotypes:

All is still,
A balmy night! and tho' the stars be dim,
Yet let us think upon the vernal showers
That gladden the green earth, and we shall find
A pleasure in the dimness of the stars.

(Wright 1987: 178)

Even the great Milton comes in for some gentle castigation, for his description of the nightingale as a 'most musical, most melancholy bird'. Coleridge's response is to say that 'in nature there is nothing melancholy'; there is only a depressed poet in the woods who, 'poor wretch! filled all things with himself, / And made all gentle sounds tell back the tale of his own sorrow'. This looks like a rejection of the 'pathetic fallacy', or the ascription of human emotions to non-humans, which is followed by the poet's wish that aspiring writers spend some time musing in the woods rather than sighing in the cities, full of 'meek sympathy' for 'Philomela's pity-pleading strains'[8] of which they have no direct experience.

The appearance is a little deceptive. It is soon clear that Coleridge is not criticizing the ascription of *any* emotion to the bird, only the idea that it is melancholy. To his imagined interlocutors William and Dorothy Wordsworth, he says:

> ... we have learnt
> A different lore: we may not thus profane
> Nature's sweet voices, always full of love
> And joyance! 'Tis the merry nightingale
> That crowds, and hurries, and precipitates
> With fast thick warble his delicious notes ...
>
> (ibid.: 179)

Moreover, Coleridge is clear that exposure to this joy is essential learning; taking his baby son Hartley outside 'when he awoke / In most distressful mood', the child becomes quiet at the sight of the moon. On this basis, he states the need for environmental education: 'I deem it wise / To make him Nature's play-mate.' Charlene Spretnak, in an essay on the genesis of ecofeminism, recalls a similar experience:

> When my daughter was about 3 days old and we were still in hospital, I wrapped her up one evening and slipped outside to a little garden in the warmth of late June. I introduced her to the pine trees and the plants and the flowers, and they to her, and finally to the pearly moon wrapped in a soft haze and to the stars.
>
> (Diamond and Orenstein 1990: 13)

Unlike Novalis, neither express a fascination with darkness and death – these are different sorts of 'hymn to the night' entirely:

> ... if that Heaven
> Should give me life, his childhood shall grow up
> Familiar with these songs, that with the night
> He may associate joy.
>
> (Wright 1987: 181)

Keats, on the other hand, does evince a wish to 'leave the world unseen' in his wonderful 'Ode to a Nightingale'. He desires to:

> Fade far away, dissolve, and quite forget
>   What thou among the leaves hast never known,
> The weariness, the fever, and the fret
>   Here, where men sit and hear each other groan.
>
> (ibid.: 277)

Envious of the easy joy of the 'light-winged Dryad of the trees', Keats joins it amid the thickets with the aid of poesy, but again finds its natural innocence and indifference provokes morbid thoughts:

> Now more than ever seems it rich to die,
>   To cease upon the midnight with no pain,
>   While thou art pouring forth thy soul abroad
>   In such an ecstasy!
>   Still wouldst thou sing, and I have ears in vain –
>   To thy high requiem become a sod.
>
> (ibid.: 278)

This, and the poem as a whole, is really a crisis of human consciousness, cast adrift in an essentially impassive world. It is dominated by blockages of the senses which throw the poet back on to the problematic resources of the individual psyche. Like Blake's tyger, Keats's nightingale does not live in a forest somewhere, as shown by the quote above which claims immortality for it. Perhaps, as Keats says, it sings the 'self-same song' that provoked Ruth 'when, sick for home, / She stood in tears amid the alien corn' in the Bible; nevertheless, it is an immortality of an exclusively literary variety. The real subject of the poem is the poet's 'sole self', and the final issue one of epistemological doubt:

> Was it a vision, or a waking dream?
>   Fled is that music: – Do I wake or sleep?
>
> (ibid.: 279)

Von Eichendorff is little known outside Germany, but his popularity there is assured through the adoption of his pleasant lyrics for folk-songs. He is also regarded as the most important German Romantic poet of nature. The nightingale appears in his poem 'Nocturne' ('Nachts') where it interrupts the grey stillness of a cloudy night. It is the only sign of life in a landscape of dripping trees and distant waters; the short lyric is dominated by the adjectives 'still' ('stille') and 'darkening' ('dunklen'). The 'nocturne' is not, as expected, the song of the nightingale, but the nocturnal song ('Nachtgesang') of the seductive, gloomy landscape itself, which at last literally 'be-wilders' the night-wandering poet. Paradoxically, the bewitchment of poet by nature is only challenged by the nightingale, its denizen, but the bird's song fails to prevent the spell of 'the night' from ensnaring the human singer. Addressing the dark itself, like Novalis, he accepts his submission:

> Under the spell you cast
> My wandering song is lost
> And like a crying-out from dreams.
>
> (Flores 1960: 111, modified translation)

Nature is once again largely a convention, devoid of concrete content, deployed in an encounter between the longing human subject and some version of 'the infinite' – God, night, the divine or the universe. In von Eichendorff, even swans can suffer – or enjoy – the longing for death which amounts to a kind of intoxication ('todestrunken')!

There is one other poet's nightingale, and the voice is unmistakable:

> Up this green woodland-ride let's softly rove,
> And list the nightingale – she dwells just here.

> (Wright 1987: 253)

This is not a poetic figure or mythological character, but a real bird, that must be approached with care for fear it will fly away. Indeed, it is not even an easy bird to find in the first instance; Clare asks us to come with him to the place to search:

> There have I hunted like a very boy,
> Creeping on hands and knees through matted thorn
> To find her nest and see her feed her young.
> And vainly did I many hours employ:
> All seemed as hidden as a thought unborn.

> (ibid.)

After these others, it is strange and refreshing to be rummaging through the undergrowth like this, not in search of 'Philomela' nor a 'Dryad of the trees', but a real bird that sings from the midst of some real hazel-bush. The direct address and detail adds to the immediacy:

> Hark! there she is as usual – let's be hush –
> For in this blackthorn-clump, if rightly guessed,
> Her curious house is hidden. Part aside
> These hazel branches in a gentle way
> And stoop right cautious 'neath the rustling boughs,
> For we will have another search to-day
> And hunt this fern-strewn thorn clump round and round.

> (ibid.: 254)

Finding the nest, Clare observes it closely, marvels at its construction and the pretty little eggs – both of which are 'curious'. His concern is not the human longing for death, nor the anguish of self-consciousness, but the survival of the bird, its curious and vulnerable nest, and the precious eggs:

> ... the old prickly thorn-bush guards them well.
> So here we'll leave them, still unknown to wrong,
> As the old woodland's legacy of song.

> (ibid.: 255)

I firmly believe this is the 'acknowledged voice of life'.

BYRON'S DARKNESS

Several important figures in British Romanticism have received short shrift here – Blake, Mary and Percy Shelley and Byron predominantly. Yet I hope to have suggested how the Romantic legacy of proto-ecological thought is both vital and ambiguous. Vital, because for the first time important writers saw and acknowledged the beauty and fragility of the natural environment, and began to formulate principles of human spiritual dependence upon the world that also feeds and clothes us. But also ambiguous: because of a widespread Romantic indebtedness to many versions of transcendental human desire; because of major historical shifts in the loci and character of environmental issues; and most importantly because Romantic nostalgia has historically supported the 'ghetto-ization' of nature, urban sprawl as the cities seep out in search of the lost world of Grasmere or Clare's Helpstone, and even, at times, extreme nationalism.

However, there may be little point in qualifying the Romantic contribution almost out of existence; we are fast depleting our limited indigenous resources of hope here in the West, and should therefore accept the Romantic offering of sympathy with and confidence in nature. We might also take from the Romantics a bleak warning of the consequences of a failure to rise to the environmental challenge. Byron's extraordinary poem 'Darkness' presents a dire vision of meteorological disaster:

> The World was void,
> The populous and the powerful was a lump,
> Seasonless, herbless, treeless, manless, lifeless –
> A lump of death – a chaos of hard clay.

> (Byron 1978: 21)

If Grasmere Vale is one extreme, a working utopia very far from any life we might presently realize, Byron's 'dream, which was not all a dream', is a similarly radical dystopia. It is, however, only as far from the present as the controllers of the nuclear arsenals choose it to be.

> The waves were dead; the tides were in their grave,
> The Moon, their mistress, had expired before;
> The winds were withered in the stagnant air,
> And the clouds perished; Darkness had no need
> Of aid from them – She was the Universe.

> (ibid.: 22)

NOTES

1 Perhaps ironically, the two German writers treated later, von Hardenberg and von Eichendorff, were both aristocrats.
2 During the Romantic period, Charles's less well-known grandfather Erasmus was very influential thanks to his agnostic popularization of contemporary biology, especially his extraordinary long poem, *The Botanic Garden*. He also advocated proto-evolutionary ideas.
3 It is important to note that this notion of a 'widening circle', and the implicit or explicit equation of animal and human rights, has been strongly criticized. The most sensitive treatment of this issue is in Midgley 1983.
4 Dorothy Wordsworth, partly on her own insistence, has remained radically marginalized in studies of this period, as have many of the other important women Romantic writers. This has extended to the subsumption within William's collected works of two of her poems – in one case without acknowledgement – and the convention whereby Dorothy is 'Dorothy' and William is 'Wordsworth'. I will not follow the latter practice.
5 See G. Garrard, 1996.
6 For a thorough discussion of this work, see Bate 1991: 41–8.
7 See R. F. Prosser, 'The Ethics of Tourism', in Cooper and Palmer 1992.
8 Coleridge is using the classical and 'poetic' name 'Philomela' ironically, to contrast the pitiful female of tradition with his modern, clear-eyed view of a joyful, male bird. Only males sing, contrary to poetic and mythological tradition.

10 Philosophy and the environmental movement

Kate Rawles

THE TROUBLE WITH FAMILIAR FURNITURE

I think it was Tolstoy who wrote that the greatest threat to life is habit. Habit, he said, destroys everything around us. By familiarizing us to the point that we no longer really see them, habit destroys our houses, landscapes, lovers and friends. Tolstoy argued that the remedy is literature – or at any rate great literature. Great literature restores our lives to us, by curing the blindness, so that we come to see what it is that surrounds us. It brings the furniture of our lives back into focus.

A similar point can be made about ideas and concepts, and about the intellectual frameworks that our ideas and concepts exist within. Concepts such as the environment, nature, humans and animals, for example, are extremely familiar, and we often take them for granted. Yet they are difficult to define, and attempts to do so will rapidly bring all sorts of complexities to light. Moreover, such concepts carry with them a whole raft of interconnected assumptions. These assumptions influence the way we think about a wide range of issues. But they often go unnoticed.

I want to suggest that the remedy here is philosophy. One role of philosophy is to restore lost sight: though what we relearn to see are patterns of ideas rather than husbands and furniture. In this chapter, I want to consider the extent to which this process is a useful one, from the point of view of those trying to bring about social and environmental change. What, if anything, can philosophy contribute to the environmental movement?

PHILOSOPHERS AND MORE USEFUL CREATURES

There's change. And there's change . . .

Philosophy and philosophers are not always highly regarded. The caricature is of philosophy as an abstract and rather irrelevant discipline carried out in ivory towers, well removed from the real world, by bright but not very sensible academics. At its worst, this – to paraphrase the inimitable

Kingsley Amis – has philosophical activity as the process of 'shining a pseudo-light on a non-problem'. It is tempting to draw a contrast between this notion of philosophizing, and actually getting things done. But philosophy, though it can come uncomfortably close to Amis's caricature (he was in fact referring to academics in general), does not have to be like this. And any straight comparison between thinkers and doers is clearly overdrawn. The environmental movement (using the phrase as an umbrella term to refer to the diverse individuals and organizations who are concerned about environmental and social issues) wants not just change but change for the better. What *counts* as change for the better needs a deal of thought. This seems especially so in the so-called postmodern era, when a range of supposed certainties has been called into question.

The attempt to figure out what would constitute desirable change clearly has a philosophical dimension to it, not least because of the need to get entangled in debate about values. I will return to this shortly. Meanwhile, here is one rather obvious way in which philosophy can be of some use. There are a number of others.

The fallacy of misplaced concreteness

First, philosophy can help prevent outbreaks of the scourge sometimes known as the fallacy of misplaced concreteness. This is the phenomenon whereby a concept is used so frequently that, after a time, everyone using it not only assumes that they know what they mean when they use it, but that everyone else means the same; and that the concept in question refers to something quite solid and uncontroversial. The phenomenon can cause trouble in a number of ways. One is that people who think they are talking about the same thing may not be. The word 'animal' is a case in point. In debates about whether animals have rights, for example, advocates often use the word to refer to a small subclass of animals, such as mammals or any animals that are capable of experiencing pleasure and pain. Those who find the concept of animal rights absurd are often using the term in its zoologically correct sense, to refer to members of the animal kingdom. The vast majority of these are not, of course, mammals. Thus, a first step in such a debate is to establish exactly which group of creatures it is whose rights are under dispute. If this is not done, the debate risks running for ever at cross-purposes.

Another risk is that of being soothed into a misplaced sense of security. The declarations of governments and institutions across the world on the subject of sustainable development would, if taken at face value, leave us assured in the belief that our environmental and social problems are on the brink of final resolution. It is well-nigh impossible to find anyone who is opposed to sustainable development. Scepticism is due here, not just in virtue of the gap between promises and implementation, but because what government ministers mean by 'sustainable development' may, for example, turn out to be light years away from what environmental campaigners mean by it. And different campaigners are themselves likely to endorse different views.

So one role for the philosopher is that of raising questions about the way in which key concepts are being used, and what is meant by them. Dispelling the fallacy of misplaced concreteness is thus useful as a sort of ground-clearing exercise, with the philosophers carefully and consistently worrying away at the foundations of our ideas and activities. Hence the popular description of philosophers as the building inspectors of thought.

Philosophical plumbing

Part of philosophical eye-opening, then, involves re-examining particular concepts. But concepts do not exist in isolation. They are embedded in a whole structure of interconnected ideas and assumptions: they are part of a wider intellectual framework which has evolved, in culture, over time. They can therefore become old and outmoded.

Mary Midgley draws an analogy with plumbing. In our houses and other buildings, a complex system of pipes, drains, taps and s-bends carries crucial stuff, like water, and underpins a whole variety of mundane but vital activities. Similarly, a complex system of interrelated ideas and concepts underpins our thoughts. Like plumbing, the conceptual structures within which we think often go unnoticed. We don't, as a rule, think about thinking, we just think. And, also like plumbing, this system can go wrong. Here, though, the analogy begins to come apart. For, as Midgley puts it, 'when the concepts we are living by function badly, they do not usually drip audibly through the ceiling or swamp the kitchen floor. They just quietly distort and obstruct our thinking' (Midgley 1992a: 139).

A further function for the philosopher is thus suggested: that of inspecting our intellectual pipes and s-bends. Part of the purpose of this is simply to bring them to our attention. We can then become aware of ways in which they are influencing our thoughts on a particular topic, and consider whether this influence is benign or otherwise.

Take animals again. To the extent that we think about our relationship with animals at all, we tend to do so within the context of an intellectual framework *that is already in place*. For example, even in the process of asking about the sort of moral obligations that humans have to other animals, we may already have made the assumption that humans are uniquely different from other animals – simply in virtue of the way in which the concepts, humans and animals, are normally understood. Such an assumption may be both correct and benign. On the other hand, it is sometimes used to support the further assumption that humans are more important than other animals, and that the welfare of non-humans is something we can therefore put on the ethical back-burner. This may turn out to be correct. But it is hardly an unbiased starting point for a debate about the relationship between humans and other animals.

Similar points can be made in the broader environmental context. Suppose that the concept, 'environment', for example, turns out, on analysis, to be widely understood to exclude humans. Suppose also that 'human' carries with it the sense of a creature somehow detached from the rest of the natural world. It would follow that, in the process of considering how our environmental problems might be resolved, we are perpetuating a particular conception of our own place in the grander scheme of things: a view of humans as standing somehow outside the environment, seeking to manage it. Again, this view of ourselves may turn out to be a felicitous one. But some have argued that it has been a contributory cause of the very problems we are trying to ameliorate. Either way, the point is that particular conceptions of our own place in the grander scheme of things are open for debate, and should not just be taken on trust.

In the case of thinking about the environment and our place in it, it is increasingly argued that our intellectual plumbing is in a bad way. The pipes are leaking: they may even be lined with lead. This has generated a debate about whether they can be fixed with the plumber's equivalent of a sticking plaster, or whether the system needs a more radical overhaul. This debate, sometimes summarized as between shallow and deep approaches to environmental issues, is an intriguing one. It may, more-over, turn out to be highly significant. If it is true that the very ways in which we tend to think and debate about environmental issues perpetuate attitudes that are themselves harmful, then 'thinking about thinking' will turn out to have practical consequences of considerable importance to the environmental movement.

VALUES

We have, then, philosophers as building inspectors, burrowing about in the foundations of our ideas, taking a critical look both at particular concepts and the intellectual framework in which they are embedded. I have tried to suggest that this sort of activity may be of use to the environmental movement in a number of quite general ways. Philosophical activity has a more specific relevance in the context of debates about values.

Environmental problems are often presented as scientific problems. Many environmental issues do, of course, confront us with questions that are essentially scientific in nature. But they also, *without exception*, raise questions of ethics or values. This is not always recognized; and the ethical or evaluative dimension to environmental problems may therefore be rather neglected. For this reason, I want now to focus on the particular branch of philosophy which deals with these matters: ethics, or moral philosophy.

First, some clarification of the terms. By 'ethics' I mean codes of conduct which prescribe what we should or should not do. I take 'values' to refer to things we aim towards, that we regard as worthwhile or good in some sense. This distinction is drawn only in a rough and ready manner. It is also worth noting that not all values are moral. Some may be aesthetic or spiritual, or they may involve matters of etiquette. Ethics and moral values are typically to do with very important aspects of human life and conduct, or very significant influences on the lives and well-being of others (as opposed to, say, issues about which fork to use first at dinner).

Moral philosophy – which has little or nothing to do with moralizing, despite its unendearing name – can be useful in a number of ways.

Identifying values

Consider air pollution from anthropogenic carbon emissions. This is an issue whose aura is thoroughly scientific. Scientific and technological questions, such as how the pollution is caused and how it might be reduced, will certainly be raised in any discussion about it. But the example also raises ethical issues, value issues and questions about our fundamental perceptions of ourselves and the world around us.

This emerges if it is asked why pollution concerns us. The answer, presumably, is because it adversely affects human welfare. This raises a straightforwardly ethical issue, to do with constraining activities which cause serious harm to others. Here also is reference to human welfare as something we value.

Other value questions embedded in this issue can be brought out by asking whether *all* pollution is unacceptable. Achieving zero pollution, even if this were technically feasible, would have extraordinary social and economic costs. The question, therefore, becomes one of balancing a reduction in pollution against other things we value. A whole herd of ethical and value issues gallop over the horizon at this point. How do we compare different values, and who should do the comparing? What sort of society, with what sort of values, do we want for ourselves and for our descendants? Who should decide these issues? By what process? and so on.

Similar issues are embedded in all environmental problems, no matter how technical they may seem on the surface. The very description of global warming, pollution, biodiversity loss and the rest as environmental *problems*, rather than environmental changes, indicates that values are involved. So an initial role for philosophy is that of uncovering these values, and bringing them to the surface so that they can be openly discussed.

If this is not done, value positions will be adopted anyway. Values lurk in the undergrowth of environmental issues whether or not they are noticed. They do not go away, either through being unnoticed or by being noticed but ignored. If they are not made explicit, they will simply be taken on implicitly and, as it were, inadvertently.

But to take on a value position inadvertently, is, in effect, to endorse someone else's values. The extent to which this constitutes a problem may depend partly on whose values they are, though issues of personal integrity are involved regardless. Our values ought, surely, to be *our* values, adopted consciously and deliberately and for reasons we could identify – even if those we have acquired unthinkingly turn out to emanate from Mother Teresa or Jonathon Porritt.

A further point is that values endorsed inadvertently may be less than ideal. This is hardly a contentious claim to make in the context of contemporary culture where, despite rather widespread reservations about the values of consumerism, it is nevertheless all too easy to be swept along with it, and extremely hard to stand back and consider the values it offers from a more critical standpoint. We can assent to a great deal, simply by taking our place in contemporary life without asking too many questions.

Biodiversity loss and hidden world views

Refocusing attention on the particular sets of values embedded in particular lifestyles and commitments is quite clearly of relevance to an activist movement that is typically rather critical of many aspects of contemporary culture. But environmental issues also raise, or perhaps should raise, very deep questions about the underlying world views or thought paradigms within which we are operating. These are also worth bringing to the surface for critical review.

Consider the issue of biodiversity loss. This is perceived as a problem rather than merely a process of change, because we consider biodiversity to be of value. Its loss at the present rate – which has been compared to previous catastrophic extinctions – has direct and indirect economic implications for us. These include lost revenue from possible medicinal and agricultural uses of plants, and the economic implications of ecosystems which are degraded and hence more vulnerable to other forms of change. In addition, biodiversity is of scientific and educational value, while some also argue that living among a rich variety of other forms of life contributes greatly to human quality of life. This may be referred to as the aesthetic and/or spiritual value of biodiversity.

These varieties of value all have their roots in human needs and interests. The value attributed to biodiversity is thus instrumental: it is a means by which human needs and interests are satisfied.[1] Insofar as it is considered that a comprehensive account of the value of biodiversity can be given in these terms, the view can be located within a general frame-work of assumptions about humans and their place in the non-human world. These include the belief that the non-human world is primarily a human resource, and may be appropriately seen and responded to as such. This assumption is a common one, and responses to environmental problems are often worked out on the basis of it.

But biodiversity loss also raises questions that challenge this assumption. Are we entitled to destroy other species and their ecosystems? Do we have ethical obligations not just to humans whose lives may be impoverished by species extinction but to the species themselves? Such questions introduce the thought that the non-human world, or some parts of it, may have value in its own right. This is sometimes described as intrinsic or inherent value. The suggestion that the non-human, natural world has intrinsic value does not, on the face of it at least, seem compatible with the view that it is essentially a human resource. Environmental issues thus raise fundamental questions about our view of the world, and our relationship to it. Again, positions will be taken, whether deliberately or by implication. And again, it seems important to examine the 'world views' that we are committed to, for reasons of personal integrity as well as the possible connections between world views and environmentally benign or malignant lifestyles.

Thinking about values

Having identified the value dimensions of environmental issues, what then? At present, even when the value dimensions of an issue are recognized, they are often dealt with rather badly. This typically happens in one of two ways. First, values may be treated as things which should be preached, in a dogmatic way. Alternatively, values may be treated as the purely subjective preferences of individuals. Both approaches rest on a fundamental misconception about what values are. Both varieties of misconception have unfortunate practical consequences.

Dogmatic values

Dogmatically delivered values are taken to have some more or less absolute source of authority standing behind them. This, whether God, or tradition, or religious texts, lends itself to an understanding of values as things which can be preached at us, and which we can acquire simply by being told what they are. All we need do is put the correct values on a list, and then learn it. One obvious problem is that of agreeing on the list. Just as difficult, at least in the present culture, is reaching consensus on which god or tradition it is that gives the list its authority. More importantly, perhaps, the authoritarian approach fails to take into account the way in which values are reached through a process of critical reflection, experience and discussion. This is a process that individuals have to go through for themselves. It cannot be short-circuited by handing out ready-made lists, at least not if values are to have an appropriate significance in people's lives, and one that endures for the right reasons.

Of course, the view of values as rooted in religion can be more or less fundamentalist and hence leave varying amounts of room for manoeuvre and debate. A religious basis for values does not need to be completely dogmatic, though in practice it often is. But in any case, if God is taken to be the source of ethics and values, a well-known conundrum arises. Are things good or right because God recognizes that they are? Or does God say that certain things are good and right because she recognizes that they *are* good and right? The first suggestion starts to look disconcertingly arbitrary, as if God could have picked anything at all. The second suggestion avoids this problem. On this view, God calls things good, not on the whim of the moment, but because they are good. But this leaves the question of what it is that makes good things good unanswered. This conundrum amounts to another reason why contemporary philosophers, whether religious or not, tend to think it a mistake to resort to an authority such as God to explain the source and nature of ethics.

Subjective values

The second view swings off to the other extreme. This view denies that values have any objective status at all, taking them instead to be purely subjective preferences, essentially the same as tastes. Indeed, this is sometimes assumed to be the *only* alternative to the notion that values are grounded in religion.

Those who take this view often contrast ethics and values with science. Science, it is said, deals with objective, universal facts. Ethics deals with judgements that are subjective and personal and hence cannot be evaluated. Unlike scientific claims, judgements about ethics or values cannot be assessed as true or false.

But science is not, of course, purely objective; nor is it value-free. As is widely acknowledged, scientific data only make sense in the context of broader theories or paradigms and these specify, among other things, what is to be valued as data, which questions are worth asking, and so on. In other words, the practice of science is shot through with values.

Values, on the other hand, are, it can be argued, not purely subjective but open to rational adjudication. A person's values may initially be acquired in a rather uncritical way from their upbringing and culture. But most people, at some stage, ask critical questions about the values they have inherited, perhaps comparing them to the values endorsed by others, and attempting to assess their merit. As part of this process it makes perfect sense to ask someone *why* they endorse a particular set of values or why they believe, say, that capital punishment or abortion is either right or wrong. Thus claims about ethics and values are open to rational adjudication because people come to have reasons for holding them. These reasons can be evaluated. They can be more or less valid; consistent or inconsistent with other beliefs; ill or well informed.

If this were not the case, we would be forced to say that the values advocated by Gandhi and by Genghis Khan are equally valid: it is just a matter of personal taste which is preferred. But there is, surely, more to it than this. We want to leave ourselves the intellectual space to say that Gandhi's values are better. After all, we do not *behave* as if values were purely subjective, especially when confronted with values that we find deeply objectionable. In a discussion about apartheid, with an apartheid advocate, for example, most of us would try to defend our own view rather than simply allow that the apartheid advocate has a different, but perfectly legitimate, one. The contention that racial hatred is wrong is not just a personal prejudice but a claim that can be argued for – and against. If values were subjective preferences, such arguments would not make sense. Preferences are, by and large, just things we have; we do not defend rationally a taste for avocados over cucumbers.

There is much more to be said here. The point to be developed and defended is that values, properly understood, are not backed by absolute authority; nor are they just tastes. Values occupy a middle ground between these two extremes. Values are acquired by individuals, through a process of critical reflection, deliberation and discussion. This will include reflection on people's particular experiences and their emotional as well as intellectual responses to them. Values are thus open to rational adjudication, at least to a certain extent. They are the kinds of thing that we can have a reasonable debate about, and some sets of values can be held to be better than others.

In the context of values, then, the philosophical tasks are: to bring to light the value dimensions of environmental problems, and thereby make them open for discussion; and to bring to light the assumptions we make about what values, *as a general phenomenon*, are like, and take a critical look at these assumptions, challenging them where necessary. Why the first might be useful seems reasonably clear. The relevance of thinking about values at this second, rather abstract, level may be less obvious. It has such relevance none the less: because the way in which values are understood in the abstract turns out to have significant practical implications in a number of areas. I will offer some brief comments about two: environmental education and environmental policy.

ENVIRONMENTAL EDUCATION

If values are treated as purely subjective preferences, environmental education, presumably, has to consist of offering students information about
issues such as deforestation, pollution and biodiversity, accompanied by
the widest variety of views as to whether and why such phenomena are
or are not undesirable. It is not at all clear how students might be
encouraged to assess these views or discriminate between them.

On the other hand, if values are treated as having the backing of
indisputable authority, this suggests (presuming, that is, that we can agree
on the authority and what it tells us) a 'revised ten commandments'
approach to teaching ethics and values in the environmental context. On
this view, environmental education would amount to little more than
learning by rote what ought to be done with regard to a range of issues,
plus, presumably, some technical discussion about how these requirements
might best be implemented (the meaning of 'best' having already been
decided). Such an approach eschews the view that values are just preferences; but it does so at the expense of personal involvement in the process
of thinking about values critically.

The alternative is to steer a middle path. This is a fine balancing act
to achieve. It involves defending a space for the perception of values as
more than subjective; while yet insisting that values are open to critical
debate. This understanding of values suggests that students be informed
about a wide range of points of view; *and* encouraged to assess their
relative strengths and weaknesses. The idea that opposing views on the
acceptability of, say, human-caused species extinction, can simply be laid
alongside one another as preferences which different people may have
or lack should be rejected. There are rights and wrongs to be found here.
But they cannot be learned by rote. There is no automatic or algorithmic
routine by which values can simply be imparted from one person to
another. Rather, environmental education should involve encouraging
students to discuss different views, assessing their merits in the light of
their own feelings and experiences as well as the strength of the rationale
that lies behind them.

Values are thus acquired, tried out, reassessed and revised as an
ongoing process. Values reached in this way can be defended by their
advocates. They are much less vulnerable to fashion-led change. And they
become the individual's own values, in a sense that does not seem to be
true of values acquired by osmosis in the course of being brought up in
a particular time and place. Moreover, the process of assessing ideas
according to their own merits rather than in virtue of personal attachment
to them makes further discussion of ideas easier and more likely. If
widespread, such an upward spiral would greatly facilitate the possibility
of cooperation between different organizations and individuals within the
environmental movement as a whole; cooperation often made difficult at

least in part precisely because of passionate and personal attachment to particular values and ideas about how these should best be put into practice.

ENVIRONMENTAL POLICY

Environmental policy provides a second context in which the way values are understood at a general level – as subjective, dogmatic or open to critical adjudication – will have practical implications of relevance to the environmental movement. More specifically, the way in which certain policy decisions are made will be strongly influenced by the particular conception of values that is held. The techniques of cost–benefit analysis and contingent valuation, for example, are implicitly committed to the view of values as purely subjective. These techniques are used to decide a wide range of environmental issues. If it is true that values are not purely subjective, the propriety of such techniques is thrown into question.

Cost–benefit analysis uses the market as a means for making decisions. Contingent valuation is a means by which goods and services for which there is no market may nevertheless be entered into the equation. Environmental goods and services, such as particular habitats, clean air or awe-inspiring scenery are good examples. A number of methods are used to attach an economic value to these things. For example, people may be asked how much they would be willing to pay for less polluted air, or to visit a nature reserve; or they may be asked how much they would be prepared to pay to prevent a favoured view from being obscured. Both contingent valuation and cost–benefit analysis are open to a multitude of trenchant criticisms. I will limit myself to three points.

First, there is no room within the various techniques of contingent valuation to establish *why* people are willing to pay a certain amount for a given environmental commodity. The reasoning behind people's answers is rendered irrelevant. An opinion about the value of a particular habitat which is held for carefully considered reasons and a subjective and arbitrary preference for a particular habitat are treated in exactly the same way. Moreover, all (similarly strong) preferences will be given equal weight. This may sound noble, but it is in fact misguided. No attention is paid to whether a given view about, say, the building of a bypass, is ill or well informed, based on years of research or overheard in a pub. Views which are vicious, bigoted or wise are all treated alike.

Second, in seeking to establish what people's preferences are about a given issue and then, as it were, simply adding them up, contingent valuation effectively takes a snapshot of people's values at a particular point in time. It thus fails to accommodate the fact that values are acquired and developed as a lifetime's process. Perhaps any attempt to ascertain the values of others would have this failing. But the method of contingent valuation, in treating values as ready-existing preferences, actually short-circuits the process of reflection and debate by which values are formed.

Third, if values are open to rational adjudication, then deciding value-laden issues in this way is simply inappropriate. It is comparable to deciding whether we should reintroduce hanging by conducting a poll, rather than by trying, through discussion and debate, to ascertain whether the reintroduction of hanging would be the right thing to do.

In this case, then, an important philosophical role is that of pointing out that certain procedures by which a wide range of environmental policy decisions are reached are effectively based on a particular view about what kinds of thing values are. If this view is, as I have suggested, mistaken, then another task will be to point this out: and to try to figure out what sort of policy-making procedures would be suggested by a more accurate understanding of the nature of values. This could readily be the topic of a paper in its own right, and I do not have the space to go into it here. But interesting work has been done on the issue, with the suggestion, for example, that institutions more like juries than calculators would provide an appropriate means for making decisions in which questions of ethics and values are involved (see, for example, Jacobs 1995).

SUMMARY – AND A FEW FINAL THOUGHTS ABOUT PHILOSOPHY

In this chapter, I have offered a sketch of philosophical activity as the process of bringing to the surface, or relearning to see, ideas, patterns of ideas, concepts and assumptions that habit and familiarity may have rendered invisible. This may sound like an abstract, intellectual exercise, rather far removed from attempts to take on, and change, environmental and social problems 'in the real world'. But this sort of activity can in fact contribute to the environmental movement in a number of ways.

First, re-examining the way in which particular concepts are being used can often be extremely helpful. The meaning of concepts like 'nature' and 'the environment', widely used and familiar as they are, may be very difficult to pin down. As a result, the same word may be used in very different ways by people who think they are discussing the same thing, but are not. Clearing up the resulting confusion is often a constructive first move in many debates about environmental matters.

Second, the philosopher can refocus attention on the structure of ideas that particular concepts are embedded within, bringing this normally hidden structure back out into the light, and asking critical questions about it. To what extent are our conceptual tools and frameworks aiding our endeavours? Or have they eroded in such a way that they actually distort our thinking and get in the way of what we are trying to achieve?

What *are* we trying to achieve? Philosophy can also be helpful in drawing attention to the value positions implicit in attempts to bring about social and environmental change, and in any discussion of environmental problems. This is done with regard both to commitment to particular sets of values; and to assumptions about what values in general are like. Such assumptions themselves influence the way in which, for example, environmental education is pursued, and the manner in which certain sorts of policy decision are reached. So thinking about values even at this rather abstract level will turn out to have surprisingly practical implications.

Taken together, these different strands of philosophical activity can be understood as attempts to look carefully and critically at the intellectual, cultural, social and political context that we all operate within, and which we perhaps more normally take for granted. Of course, we are all creatures of our time and place and no doubt we can never completely free ourselves from the assumptions of our age. We can only stand back so far. But we can stand back a bit, and often this is far enough to see a range of problems with our current ways of doing and thinking. Hence the frequent calls, within environmental philosophy and the broader environmental movement, for a paradigm shift, or a change in 'world view'.

The more activist side of the environmental movement will not necessarily have time to stand back and think about what they are doing and why, and how their activities fit into a bigger picture. Critical reflection takes time and distance which the pressure of immediate tasks often does not allow. But this critical reflection is crucial, especially if aspects of the bigger picture are held to be flawed. It is crucial that someone is doing it. Whether this is philosophically inclined activists or actively inclined philosophers does not really matter, so long as the philosophical reflection and the activism do not become completely divorced from one another, so that both carry on in their own spheres with no communication between the two.

Precisely this, of course, can be a genuine problem, and it raises a serious complaint. Philosophy as practised in academic institutions is not always encouraged to apply itself to real-world issues. There persists a bias against applied philosophy in academic circles, despite the growing concern that universities should be able to explain the value of their activities to a broad constituency. This is reinforced in research-assessment exercises which may not reward work that is published in places not considered central to the particular specialized discipline under assessment.

But reasonable reservations can be expressed, not just about philosophy as practised in the peculiar academic context in which many of us try to work, but about philosophy as a general discipline. Jonathon Porritt once argued that, with a concept like sustainable development, the priority is to get it widely used, to get it out and about in the world, as a familiar and common notion. Then, perhaps, we can settle down to debate what, precisely, it should mean. The philosophical concern with defining terms before they are let loose gets these priorities the wrong way around. This can be broadened into a more general complaint. Sometimes, concepts and intellectual structures are working perfectly well. In this case, sustained critical analysis of them may not only amount to a waste of time and effort, but might actually be harmful.

This criticism shows philosophers to be more like vandals than building inspectors, and I think that there is always a genuine danger that philosophical activity can become destructive in this way. But the appropriate response, surely, is to keep a watchful eye on philosophical activity rather than to jettison it altogether. A second sort of complaint may be made by those at the more active end of the environmental spectrum. It's all very well for academics, this grievance goes, but we are in the real world, where there are all sorts of parameters and constraints on what we do. To a large extent, we just have to get on with it.

For example, environmental philosophers are often highly critical of the use of cost–benefit analysis, for a number of reasons. But those on the ground may point out that, if they do not use cost–benefit analysis, their concerns will simply be ignored. For example, the Royal Society for the Protection of Birds is sympathetic to criticisms of cost–benefit analysis – but argues that if it does not try to put an economic value on bird species when their habitats are threatened by, for example, road development proposals, then the 'bird factor' is not counted at all. It will just be left out of the equation altogether.

It is hard not to sympathize with this. But two further points should be made. First, it may well be true that, in the short term, the RSPB and others do have to go along with the system; do have to get on with the job. This, however, is perfectly compatible with the claim that *somebody* should be thinking critically about what the job is, exactly, and how the system in which we are getting on with it might usefully be changed. In the long term, and this is the more important point, it is precisely part of the role of philosophy, at least as I understand it, to try to change the parameters; to change the real world. The claim, well that is just the way the world is, is often a cover-up in which social or environmental ills which could be done away with are presented as inevitable, as part of reality. Social inequality is a good example of this. Philosophy can and should contribute its skills to the attempt to challenge the assumptions of the 'real world' and try to create a better one.

This brings me to my final comment. There is a double aspect to this kind of work. So far, I have been concentrating on the careful, critical, deconstructive, analytical aspect of philosophical activity: the building inspector work. This work is important. Even if we are in the business of creating something new, the problems of the old need to be figured out and the rubbish, as it were, cleared away. But if we think that we need to do more than look critically at our current values and our current conceptual schemes, if we decide that we actually need to develop new and better ones, then there is constructive, creative and even visionary work to be done as well. We need plumbers who can dream up a whole new system. This creative work is crucial in the environmental movement as much if not more so than anywhere else. And it is work to which philosophy can contribute alongside many other subjects – including science, and literature. Or at any rate, great literature.

NOTE

1 Some would suggest that it is not entirely accurate to characterize aesthetic value as a variety of instrumental value, as it is the thing itself that is valued, rather than something it gains for us. But it is still a kind of value that is rooted in human needs and interests. (For discussion of this issue, see chapter 8 of this volume.)

11 Spiritual ideas, environmental concerns and educational practice

Joy A. Palmer

So far this volume has explored the attitudes towards the environment implicit or explicit in the world's major religions and in pantheism; and has considered various conceptions which accord intrinsic value and meaning to nature. Attention now turns to a focus on the implications for practice of religious and spiritual ideas and attitudes towards the natural world – commencing here with the field of education, then moving on to examine implications in the fields of personal development and environmental policy.

This chapter sets out to describe links between spiritual ideas and attitudes held by individuals and the development of environmental awareness and concern; then to discuss the implications of these links for policy and practice in environmental education.

SPIRITUAL DEVELOPMENT

First, by way of introduction, it is useful to explain what might be understood by the term 'spiritual development' in the context of a formal education service. An overview from Britain is used as an example. (For those wishing to pursue a broader and more in-depth study of meanings and interpretations of spiritual education than can be provided here, Carr (1996) provides an overview of conceptions in the field.)

The *Education Reform Act* (1988) sets education within the context of the spiritual, moral, cultural, mental and physical development of pupils and of society; a set of dimensions which permeate the whole curriculum and ethos of schools. Thus it refers to a 'spiritual' dimension of human existence which applies to all pupils and which should be addressed throughout the curriculum, not just in the teaching of religious education. A discussion paper prepared by the National Curriculum Council on the subject of spiritual and moral development (NCC 1993) makes clear that spiritual development is not synonymous with the development of religious beliefs or conversion to a particular faith. Instead, it sees the term as applying to something fundamental in the human condition which is not necessarily experienced through the physical senses

and/or expressed through everyday language. It has to do with relationships with other people and, for believers, with God; with the universal
search for individual identity; with our responses to challenging experiences, and encounters with good and evil; and with the search for meaning
and purpose in life and for values by which to live. There are many
aspects of spiritual development:

- *Beliefs.* The development of personal beliefs, including religious beliefs;
  an appreciation that people have individual and shared beliefs on
  which they base their lives; a developing understanding of how beliefs
  contribute to personal identity.
- *A sense of awe, wonder and mystery.* Being inspired by the natural
  world, mystery or human achievement.
- *Experiencing feelings of transcendence.* Feelings which may give rise to
  belief in the existence of a divine being, or the belief that one's inner
  resources provide the ability to rise above everyday experiences.
- *Search for meaning and purpose.* Asking 'why me?' at times of hardship
  or suffering; reflecting on the origins and purpose of life; responding
  to challenging experiences of life such as beauty, suffering and death.
- *Self-knowledge.* An awareness of oneself in terms of thoughts, feelings,
  emotions, responsibilities and experiences; a growing understanding
  and acceptance of individual identity, the development of self-respect.
- *Relationships.* Recognizing and valuing the worth of each individual;
  developing a sense of community; the ability to build up relationships
  with others.
- *Creativity.* Expressing innermost thoughts and feelings through, for
  example, art, music, literature and crafts; exercising the imagination,
  inspiration, intuition and insight.
- *Feelings and emotions.* The sense of being moved by beauty and kindness; hurt by injustice or aggression; a growing awareness of when it
  is important to control emotions and feelings; and how to learn to use
  such feelings as a source of growth.

(NCC 1993: 2–3)

The Office for Standards in Education describes the scope of spiritual
development, as stated with the Framework for Inspection of Schools as:

Spiritual development relates to that aspect of inner life through which
pupils acquire insights into their personal existence which are of
enduring worth. It is characterised by reflection, the attribution of
meaning to experience, valuing a non-material dimension to life and
intimations of an enduring reality.

(OFSTED 1994: 8)

Of course, spiritual development is a very individual matter; people differ
in their interpretations of its components and in the range and depth of

meanings they ascribe to them. Many find meaning and explanations in
the teachings of a particular religion; others solely in the course of
everyday experiences. Yet the notion that people develop spiritually suggests that it is an aspect of an individual's life in which progress and
changes can be made, and thus that it is an integral and important element
of education. National Curriculum guidance does not advocate a model of
linear progression in spiritual development, but identifies certain steps
which may be achieved, viz:

- recognizing the existence of others as independent from oneself;
- becoming aware of and reflecting on experience;
- questioning and exploring the meaning of experience;
- understanding and evaluating a range of possible responses and interpretations;
- developing personal views and insights;
- applying the insights gained with increasing degrees of perception to
  one's own life.

(NCC 1993: 4)

Spiritual development is fundamental to other areas of learning since it
incorporates the curiosity, the use of the imagination, insight and intuition,
and the desire to question and to explore the meaning of experience,
which underpin motivation to learn in general.

THE ROLE OF SPIRITUAL IDEAS AND ATTITUDES IN THE DEVELOPMENT OF ENVIRONMENTAL AWARENESS AND CONCERN

A number of on-going empirical research studies provide fascinating and
perhaps influential data which confirm the hypothesis that spiritual ideas,
attitudes and experiences held or encountered by individuals may strongly
influence their development of awareness of and concern for the environment, and the subsequent adoption of pro-environmental adult behaviour.

In the limited space available, three studies will be referred to that the
present author is engaged in; the first in a reasonable degree of detail
and the others as briefer but relevant citations for corroboration of
findings.

The first research study, on the development of concern for the environment and influences and experiences affecting the pro-environmental
behaviour of educators (Palmer 1993; Palmer and Suggate 1996), examines the relative importance of various categories of influence and
formative life experiences on the development of environmental educators' knowledge of and concern for the environment. The motivation
for this study was the belief that if a fundamental aim of education is
to help children and students understand, appreciate and care for the

environment, then those responsible for this area of the curriculum should know the types of learning experience that help to develop active and informed minds.

The study was distributed in the first instance to the membership of the National Association for Environmental Education in the UK and later to various international destinations, as discussed below. Subjects were asked to provide their approximate age, gender, details of their demonstration of practical concern for the environment and an autobiographical statement identifying those experiences and formative influences that led to this concern. The participants were also asked to state what they considered to be their most significant life experiences and to write a statement indicating which, if any, of the years of their lives were particularly memorable in the development of positive attitudes towards the environment. As the outline and proformas gave only the aims and purposes of the research the participants were able to provide original responses unbiased by any examples. We aimed to confirm the sample as a group of active and informed citizens: that is, those who know about and care for the environment in their adult life. A list of seven possible activities relating to pro-environmental behaviours was provided, and the subjects were asked to indicate those in which they are regularly engaged.

In the UK sample there were 232 responses, 102 from male subjects and 130 from female subjects. Of the respondents 55 were in the under 30 age group, 124 in the 30–50 years group, and 53 in the over 50 group. The responses to the questions concerning practical activities certainly confirmed a commitment to environmental concerns. Over 90 per cent of the sample participated in some categories and 62 per cent or more participated in all but one category.

The autobiographical statements giving details of formative influences and significant life experiences leading to a commitment to environmental concerns were analysed, and the results were coded into categories of response. Thirty preliminary categories were derived in the initial analysis, which were then refined into thirteen categories incorporating various sub-categories from the original list.

The number of subjects identifying with each major category of response is shown in Table 1.

The category 'outdoors' includes three substantial subcategories: childhood outdoors (97 respondents), outdoor activities (90 respondents) and wilderness/solitude (24 respondents). 'Education/courses' refers to two subcategories: higher education or other courses taken as an adult (85 respondents) and school courses (51 respondents).

Much fuller details of the data analysis and conclusions drawn will be found elsewhere – of both the initial analysis (Palmer 1993) and of a more fine-grained analysis which looks not only at patterns of influence across the whole sample, but by age group (Palmer and Suggate 1996).

Table 1: Formative influences on environmental concern

| | Respondents | |
Category of influence	No.	%
Outdoors	211	91
Education/Courses	136	59
Parents/Close relatives	88	38
Organizations	83	36
TV/Media	53	23
Friends/Other individuals	49	21
Travel abroad	44	19
Disasters/Negative issues	41	18
Books	35	15
Becoming a parent	20	9
Keeping pets/animals	14	6
Religion/God	13	6
Others	35	15

The relevance of this work here lies in its illumination of the apparently very strong influence of spiritual ideas, attitudes and experiences on people's environmental thinking as well as the influence, for some, of religious beliefs. Thirteen subjects (6 per cent) wrote specifically about the key influence of God or a religious faith; for example:

> Son of a farming family . . . brought up on the North Essex sea coast . . . fully influenced by the work of the great architect of the universe . . . A strong feeling of the Creator's presence in the lane, field and garden . . . being born of parents who acknowledged God as the creator of the living world.

Both the initial and fine-grained analyses of the data show that whilst, as one would expect, such things as education courses, parents, other close relatives, friends, books, TV, media, the impact of environmental disasters, travel, and so on all play a significant role in promoting environmental awareness and concern, the single most important influence overall is childhood experiences 'outdoors' – in the natural world. The ninety-seven references to childhood outdoors include accounts of growing up in a rural area, holidays in the country, and numerous outdoor pastimes. Many of these accounts are very rich in references to aspects of spiritual thinking and experience:

> A strong natural affinity to living things, felt in early childhood . . . I must have been born in a hedgerow . . . soon realized that I got a better sermon from a beech tree than a bishop.

So, too, are many of the responses in the other highly significant 'outdoors' categories – outdoor activities (90 responses) and wilderness/solitude (24 responses). Many subjects mentioned feelings of 'awe and wonder'; of 'mystery' and 'transcendence' when describing time spent in the natural world – either living in a rural environment or engaging in activities such as walking, camping, bird-watching, practical conservation tasks, gardening and farming. Twenty-four subjects referred specifically to the spiritual influence of remote places, open space, and the experience of solitude or freedom:

> Allowing time to become absorbed in just watching... opening my eyes and mind to much greater things outside ourselves and our comprehension.

Complex and often passionate personal accounts such as those written for this study are inevitably highly qualitative. They do not, and perhaps should not, lend themselves to statistical analysis – yet it is estimated (and substantiated with inter-judge reliability) that over 90 per cent of the respondents referred at some point in their life stories to spiritual ideas and experiences as key influences on their thinking. As already suggested, many of these references came at the point of describing the influence of being in the natural world or of a religious upbringing; but various other categories of response, notably the influence of parents, close relatives and friends, portrayed clearly spiritual dimensions.

Following on from this analysis of autobiographical data obtained in the UK, the study has now been extended to incorporate parallel data from a range of international locations including Australia, Canada, the USA, Greece, Slovenia, Sri Lanka, Uganda and Hong Kong. On-going analysis suggests a reinforcement of the crucial importance of early childhood experiences in the natural world, and of experiences in the outdoor world in general, irrespective of fascinating cultural differences. Preliminary analysis of international data also suggests powerful reinforcement of the lasting impact of spiritual and religious ideas, beliefs and experiences on people's thinking in relation to the environment. This is illustrated here with some appropriate quotes from the Sri Lankan data, which provide an interesting cultural contrast to the UK accounts:

> When I was about twelve years old I came to understand that human beings are destroying our beautiful world... I used to listen to a great Buddhist priest in Colombo. His preaching contributed much in inculcating positive attitudes towards flora and fauna and the whole environment... the religion, teachers and knowledge I got in my school influenced me a great deal.

> During a severe drought about twenty-three years ago when all the wells ran dry... my family had contracted hepatitis and my two-year-old daughter was infected... the sight of the hillsides being cut down and paddy fields filled up for buildings... the increasing number of sawmills in our village... they were tremendous influences, and I

thought of my childhood ... where trees and the protection of animals and plant life are so important.

The present environment is artificially created among a concrete forest where there was a rain forest in which my forefathers used to live so contented healthier life than today.

When I was schooling one fine morning, my English teacher accompanied me to her residence. I was wandering here and there in the garden and I came to a little chamber where I began to feel that I was in a smooth cool, green heaven ... I felt that I was moving with my close beloved ones.

Two other on-going research projects, which can only be referred to with brevity here, provide further evidence of the importance of spiritual ideas, attitudes and values in the development of people's awareness of and concern for the environment. First, a project on 'The Global Environment and the Expanding Moral Circle' (Cooper and Palmer publications in preparation) aims to investigate recent changes in attitudes and feelings of responsibility towards the environment, animals and future generations. As part of the project, 182 individual subjects from the local community filled in questionnaires which probed their views and attitudes relating to the research agenda. In response to the question, 'What would you identify as the single most important influence or experience which has affected your attitude to our responsibility towards a) animals? and b) the environment?', 62 subjects (34 per cent) provided a response for animals and 37 subjects (20 per cent) a response for environment that could be classed as being in the religious/spiritual domain.

Another question asked subjects to provide an assessment on a scale from 0 (not at all important) to 5 (very important), of the influence of certain things (media images, TV documentaries, books, intellectual argument, parents, friends, travel, etc.) on their attitude to responsibility towards animals and the environment. Figures on this assessment relating to spiritual domain experiences (personal experience in or with nature, etc.) are as follows:

Assessment	0	1	2	3	4	5	
	1	16	23	35	39	56	*Number of respondents*

Such experiences ranked third out of fourteen categories of influence, after TV documentaries and media images.

Second, a project on 'Subject and Community Knowledge in Environmental Education' (Palmer publications in preparation) aims to investigate various forms of knowledge and awareness of environmental issues possessed by fifty undergraduate students of education. Each subject has been interviewed individually and asked questions relating to his or her own level of understanding of environmental issues, views on what an environmental education programme for school pupils should contain, how knowledge about the environment should be conveyed to pupils, experiences that have helped prepare the student for effective teaching in this area, and sources of environmental knowledge and ideas.

Once again, the data reveal a rich variety of responses, including many that would be classified in the affective/spiritual domain. For example, the lead question, 'If you were designing an environmental education programme in a primary school, what would you consider to be essential core content of it?', led to some straightforward responses such as 'Waste... picking up litter... pollution... global warming, etc.', and some far more reflective ones:

> An appreciation of who we are with regards to the natural environment... I suppose it's more of a philosophical stance than a scientific one... to understand that we're part of a whole, not governors or lords... just one link in a large chain... and to realize that although we're just one link in that chain, the effect we have on the environment is disproportionate.

Responses to questions probing sources of students' knowledge were divided into two main groups: 'formal' sources including school, higher education courses, TV media, and so on; and 'community or informal' sources leading to knowledge acquired by living and interacting with people in a community or locality and so absorbing facts, ideas and attitudes from these experiences. Nineteen respondents, 38 per cent of the sample, said that community or informal sources had been the most important for their own learning and motivation to teach environmental education. A further twelve respondents (24 per cent) said that they gave equal weighting to formal and informal sources. Responses revealed a powerful sense of the influence of such things as insight and intuition; being moved by beauty and example; being hurt by injustice; being aware of and reflecting on the meaning of experience, and so on – key facts of the individual's spiritual development:

> Cruelty to animals... it was a sort of gut reaction... you see somebody being cruel and you think 'That's wrong!'... I think those are the sorts of things you pick up... When you see someone treating animals well, then you think 'That's the way it should be.'

> The children have the seeds... they see them grow... they nurture them, and there are tears if something happens to them... it gives them the feeling that they have brought something into the world... and they know the pain when it goes. So it makes them more aware of

other people ... that they won't go out and harm anyone ... they learn to nurture each other.

THE INFLUENCE OF ENVIRONMENTAL EDUCATION

The research studies cited above provide firm evidence of the link between spiritual ideas, attitudes and experiences, and the development of environmental understanding. They also provide valuable insights into the role and importance of environmental education. In the first study, on formative influences, 'education/courses' was the second largest major category of response, after experiences 'outdoors'. 'Education/courses' includes two subcategories: higher education or other courses taken as an adult (85 respondents) and school courses (51 respondents). The apparent impact of secondary and higher education courses on environmental understanding is very encouraging – of the seven subjects who cited education as being the single most important influence on their environmental thinking, five were writing about degree level courses and two about A-level courses. Of the 51 respondents citing school-based work as a significant influence, 38 referred to A-level courses and related fieldwork. Perhaps the most disturbing aspects of the data are that twenty-three individuals chose to report that school programmes had had no influence upon them at all, and there was not one reference to a school course below A level as a single most important influence. All of the many accounts of the importance of childhood days talked of such things as experiences outdoors, influence of parents, friends, the media and so on, with very few references to primary school lessons.

In the second research study on changing attitudes, only 9 out of 182 subjects (5 per cent) cited education as the single most important influence affecting their attitudes and sense of responsibility towards animals, and 25 subjects (13 per cent) cited it as the single most important influence affecting attitudes and responsibility towards the environment in general. Figures for the assessment on a scale from 0 (not at all important) to 5 (very important) of the influence of education courses on attitude to responsibility towards animals and the environment are as follows:

Assessment	0	1	2	3	4	5	
School level courses	9	43	34	34	32	18	*Number of*
Higher level courses	6	36	28	36	34	30	*respondents*

Analysis of this assessment of influence question as a whole shows TV documentaries to be the most important influence overall, followed by media images, personal experience with animals and nature, nature and wildlife films, and intellectual argument. Higher level education courses came sixth in ranking, and school level courses ninth – out of fourteen categories of influence.

Finally, in the third study on students' subject and community knowledge, only nineteen of the subjects (38 per cent) said that formal education (including books and TV), as opposed to knowledge acquired informally by living and interacting within a community, was the most significant way in which they had acquired environmental understanding and concern. This seems a surprisingly low number, given the background of subjects in this particular sample – all are university undergraduates who have been extremely successful in the realm of formal education.

The overall picture emerging from these various projects is most interesting: it suggests that programmes of environmental education in the formal education service are indeed playing an important role, both in the development of people's knowledge and understanding of the environment and in their formulation of attitudes and feelings of responsibility towards it. Yet the influence of environmental education, even where well-structured programmes are provided, is certainly not as dominant or successful as perhaps it ought to or could be. For many individuals, ideas and personal experiences in the spiritual and religious domains have had as great or even greater impact than formal education upon their development of environmental awareness, concern and responsibility. Furthermore, the research suggests a curious yet distinct separation of these two spheres of influence on people's lives.

ENVIRONMENTAL EDUCATION: PERSPECTIVES, POSSIBILITIES AND PROBLEMS

Let us now turn, then, to look briefly at some past and present perspectives on environmental education, which may provide some degree of explanation of the findings described. Environmental education has become widely recognized for its significance over a period of around thirty years. Disinger (1983) writes that the term 'environmental education' was first used in 1948 at a meeting in Paris of the International Union for the Conservation of Nature and Natural Resources (IUCN). Yet it is in the more recent past of the late 1960s onwards that environmental education has been debated and promoted on a global scene. A landmark international IUCN/UNESCO meeting on Environmental Education in the School Curriculum held in Nevada in 1970 arrived at a definition that was widely adopted around the world:

Environmental education is the process of recognizing values and clarifying concepts in order to develop skills and attitudes necessary to understand and appreciate the interrelatedness among man, his culture and bio-physical surroundings.

(IUCN 1970)

In 1972 a United Nations Conference on the Human Environment was held in Stockholm: the first world meeting on the state of the environment. This endorsed the fundamental importance of environmental education and was followed by the establishment of the United Nations Environment Programme (UNEP). In 1975 UNEP and UNESCO founded an International Environmental Education Programme (IEEP). This programme was responsible for a key International Workshop on Environmental Education, held in Belgrade in 1975, which produced a set of principles of environmental education in a document entitled *The Belgrade Charter – A Global Framework for Environmental Education*. This charter emphasized that a world-wide environmental education programme was essential for the resolution of environmental issues, and led to the staging of the First Intergovernmental Conference on Environmental Education, attended by sixty-six UNESCO member states, in Tbilisi in 1977. A declaration from this conference, based on the principles outlined in the Belgrade Charter, established a framework for an international consensus on environmental education. It defined objectives in terms of awareness, knowledge, attitudes, skills and participation; and laid down a set of guiding principles. A key message of the Tbilisi Declaration (UNESCO 1978) was that environmental education should further the development of conduct compatible with the preservation and improvement of the environment. From that time onwards, environmental education held enhanced and recognized status throughout the world. The Commission of the European Communities established an Environmental Education Network Project in 1977, and in 1980 a major international document, the *World Conservation Strategy* (*WCS*) was produced by IUCN, UNEP and the Worldwide Fund for Nature. The key concept introduced in the *WCS* is that of sustainable development, the idea that conservation and development can be mutually interdependent. A chapter of the *WCS* was devoted to environmental education, claiming that:

> A new ethic, embracing plants and animals as well as people, is required for human societies to live in harmony with the natural world on which they depend for survival and well-being ... it is the long-term task of environmental education to foster or reinforce attitudes and behaviour ... compatible with this ethic.
>
> (IUCN, UNEP, WWF 1980)

The *World Conservation Strategy*'s message was significantly reinforced and extended in 1987 by the report of the World Commission on Environment and Development, *Our Common Future* (Brundtland Report). This report articulates in detail the global need to reconcile environment with development and stresses the fundamental role of education as a tool in achieving this. A UNESCO/UNEP conference held in the same year endorsed both the recommendations for action laid down in the Brundtland Report and the principles of environmental education identified in Tbilisi a decade earlier.

The European Community continued to be active in the recognition of the importance of environmental education. In May 1988 a resolution was passed by its Council of Ministers which identified it as an entitlement for every school pupil in the Community, and encouraged all governments to review provision for it; and in 1992 the resolution was reinforced with a statement to the effect that environmental education should be an integral and essential part of every European citizen's upbringing.

If environmental education programmes are to address the complexities of sustainable development, and issues of quality of life, equality and justice, then a defined core content of knowledge and understanding about the environment will be necessary. Learners will need a context in which to develop conceptual understanding. Various attempts have been made at international and national levels to arrive at a consensus of a core content of knowledge for environmental education programmes, usually reflecting the principles and recommendations of Belgrade and Tbilisi. In the UK, following the *Education Reform Act* (1988), environmental education became an officially recognized cross-curricular theme of the National Curriculum, that is, an important though non-statutory cross-curricular element of the whole curriculum. As such, it had an agreed definition and content, as published in *Curriculum Guidance 7: Environmental Education* (NCC 1990). These guidelines outline the core aims and objectives to be covered in the design and implementation of programmes of work in environmental education. These include objectives expressed in terms of areas of knowledge and understanding to be pursued through this cross-curricular theme:

As a basis for making informed judgements about the environment pupils should develop knowledge and understanding of:

- the natural processes which take place in the environment;
- the impact of human activities on the environment;
- different environments, both past and present;
- environmental issues such as the greenhouse effect, acid rain, air pollution;
- local, national and international legislative controls to protect and manage the environment; how policies and decisions are made about the environment;
- the environmental interdependence of individuals, groups, communities and nations;
- how human lives and livelihoods are dependent on the environment;
- the conflicts which can arise about environmental issues;
- how the environment has been affected by past decisions and actions;
- the importance of planning, design and aesthetic considerations;
- the importance of effective action to protect and manage the environment.

(NCC 1990: 4)

The same document suggests that a basic knowledge and understanding of the environment can be developed through the following topics: climate; soils, rocks and minerals; water; materials and resources, including energy; plants and animals; people and their communities; buildings, industrialization and waste. It was envisaged at the outset of National Curriculum implementation that generally these topics will be taught through core and foundation subjects, and in particular through the attainment targets and programmes of study for science, technology, geography and history. The revised statutory orders for the National Curriculum for Schools (1995) made no reference to reinforcing the status and importance of its cross-curricular themes. Instead, a framework for teaching essential aspects of knowledge and understanding relating to the environment is subsumed within the orders for geography and science education. There are also references to it within the technology curriculum and what might be described as 'opportunities' in other subjects.

Meanwhile, on the global scene, 1991 was marked by the launch of *Caring for the Earth* (IUCN, UNEP, WWF 1991), a completely revised and updated version of the *World Conservation Strategy* which sets out nine principles of sustainable living and emphasizes the crucial tools of education and training in achieving them. A year later, the largest conference ever to be organized on environmental matters, the United Nations Conference on Environment and Development (UNCED), was staged in Rio de Janeiro, Brazil. The major document deriving from this, *Agenda 21*, devotes a chapter to education and training which stresses the critical importance of education in the promotion of sustainable development. It recommends that all governments integrate 'environment and development as a cross-cutting issue into education at all levels within three years' (UNCED 1992).

Thus it is evident that over the past three decades, pleas and recommendations have been made at a number of substantial gatherings and in widely acclaimed documents for the translation of global definitions, objectives and principles into policies, programmes and resources at national and community levels. Alongside this upsurge of interest and guidance in environmental education, numerous other related 'educations' have found their ways on to the stage: development education, global education, futures education and human rights education – all have their place in the global agenda for the promotion of pro-environmental behaviours and the improvement of the quality of human life. The development and existence of these various interlinked approaches to environmentalism are an indication of the tremendous energy that has been, and indeed still is being, devoted within the general field of 'education' to the promotion of planned processes which enable participants to explore and understand the environment, to engage in sustainable activities and to take action to make the world a better place for all forms of life.

Yet have we actually progressed beyond the rhetoric of Belgrade, Tbilisi, Brundtland and Rio? Have we translated the numerous lists of objectives, principles and recommendations into appropriate practice? Have we really moved forward in the past thirty years in terms of the real impact and success of environmental education? (Which, in the context of this chapter, I interpret as meaning that it contributes significantly to the development of people's environmental understanding, awareness and concern; that it encourages active participation in resolving environmental problems and helps people acquire the commitment and skills needed both to appreciate and to protect the environment.) Developments in the philosophy, policies and practice of the field through time have been complex. In very generalized terms they have transformed the dominant view of good practice in environmental education from one of teaching *about* nature, with 'show-and-tell' techniques, of the early 1970s; to one of teaching *through* fieldwork in the 1980s; to one of action research, and pupil-led problem-solving fieldwork *in* the environment in the 1990s (often accepted as good practice yet rarely practised because of time and resource constraints).

Such developments in dominant trends are to be welcomed. Yet while they may be moving environmental teaching and learning in the right direction, it would seem that in global terms, formal courses of environmental education appear not to be achieving the success they deserve, despite the tremendous energy that has been channelled into their design and development – a view substantiated by the research discussed earlier in this chapter.

The explanation for this is inevitably complex, with reasons ranging from the philosophical and ideological to the purely practical. Let us consider a sample 'range' of issues.

In the first instance, there exist extensive contradictions between environmental education and accepted classroom practices. For example, the guiding principles of the Tbilisi Declaration, the messages contained within the *WCS*, *Our Common Future*, *Agenda 21* and the aims and characteristics of environmental education set out in national guidelines such as *Curriculum Guidance 7* in England and Wales and *Learning for Life* (Scottish Office 1993) in Scotland, all establish particular kinds of curriculum and pedagogical practice as being necessary to achieve the stated goals. They suggest that teaching and learning should be cooperative processes of enquiry into environmental issues, leading to taking of appropriate actions. In other words, learners are working individually and collectively towards developing environmental understanding and awareness and the resolution of on-going environmental problems. This approach requires environmental education to be interdisciplinary with a focus on real complex problems; yet, in reality, school curricula tend to be discipline-based, with an emphasis on abstract, theoretical problems. Furthermore, all the guidance suggests that learning in environmental education should be holistic, and cooperative, involving active-thinking and the generation of ideas and knowledge; whereas school-based learning places emphasis on individual learning and a focus on the assimilation of knowledge and the thinking of others.

Second, despite the rhetoric of the 'grand' statements on environmental education that makes pleas for it to 'encourage pupils to examine and interpret the environment from a variety of perspectives – physical, geographical, biological, sociological, economic, political, technological, historical, aesthetic, ethical and spiritual' (NCC 1990), in reality it remains to a large extent grounded in the scientific domain. The dominant approaches to environmental education have traditionally been, and still are, too narrowly focused on the need to communicate information about problems in order to bring about a change in people's behaviours. The field of environmental concerns or the crisis the Earth is facing has tended to be treated as a series of measurable and controllable issues such as pollution, destruction of the ozone layer, non-renewable-resource destruction, energy and conservation matters. This dominant view often excludes social, political and economic concerns.

A question asked of the fifty university undergraduates in the 'Subject and Community Knowledge' research probed what they consider to be the main issue affecting our planet at the present time. The most commonly mentioned issues were ozone layer depletion/global warming (36 references), rain forest issues/deforestation (31 references) and pollution (31 references). When asked if they could suggest the single most significant issue, forty-five out of the fifty subjects provided answers, forty of which derived from the scientific domain or were expressed primarily in terms of scientific/measurable impact upon the planet. The four most commonly cited key problem impacts were ozone layer/global warming (11 respondents), pollution (9 respondents), deforestation (6 respondents) and human population increase (expressed in mathematical, measurable impacts terms, 6 respondents). Only five out of the fifty subjects singled out an issue with an alternative focus, for example, economic, political, social, ethical. Two talked of disproportionate distribution of wealth between the West and the Third World; one of famine and acute human suffering; and two said that the single most important issue affecting the world today is the attitude of people towards the environment. These responses came from students who had recently completed school courses with success, including whatever aspects of environmental study had been incorporated in their curriculum.

Clearly their views reflect the dominant approach to environmental education which is grounded in the scientific domain and reflects the scientific world view that has dominated Western thinking as refined in the nineteenth-century thinking by Comte, Mill and Marx, and carried into the twentieth century by philosophers such as Bertrand Russell and the logical positivists (or logical empiricists) of the Vienna Circle. Despite an anti-positivist tradition, the scientific world view continues to influence approaches to various academic fields such as quantitative social science including the field of formal education.

Positivistic approaches are characterized by a basically applied science approach to educational inquiry, seeking to apply standards and methods of the natural sciences to problems and issues in education.

Any reading of the major journals and research publications in environmental education suggests that empirical enquiry in the field is also predominantly based on an applied science model as used in other areas of education, including science education. This argument is illustrated and expanded by Palmer (1998) and in a monograph on research in environmental education (Robottom and Hart 1993) where it is claimed that 'much of the research in environmental education takes the form of ascertaining the congruence between outcomes and assured goals and seeking empirically (objectively) to derive generalisations (theory) and hence legitimate scientific knowledge.' Within this paradigm, the problem of improvement of environmental education continues to be seen as a matter of identifying and controlling variables associated with such goals as responsible environmental behaviour. Indeed, a dominant focus in environmental education research in the 1980s was on the prediction and control of the variables that are believed to be the critical determinants of pro-environmental behaviour. (See, for example, Hines, Hungerford and Tomera 1986 and Ramsey and Hungerford 1989.) Hungerford (1983) identifies a number of challenges for environmental education, including 'operationalizing environmental literacy: selecting appropriate goals'; claiming that educational research has been only partly effective in identifying those 'variables which influence behaviour'.

Although researchers have investigated such variables as environmental sensitivity, knowledge of issues, beliefs, values, focus of control, and a number of population demographics, in none of the studies conducted thus far have they been able to account for a substantial amount of variance associated with behaviour in a manner that is generalizable to large segments of the population overall (Hungerford 1983).

This comment reflects the influential applied science or technocratic view of educational research, where the educational context is seen as a system made up of discrete variables which influence behaviour in a 'generalizable way'. These and other examples of research and practice (see, for example, Robottom 1987, Palmer 1998) show that the organization of environmental education, both its infrastructure and strategies for change, is predominantly hierarchical and technocratic in character. 'This form of organisation presupposes that curriculum materials, if properly prepared, can be effective change agents; that "experts" can bring pre-existing, positive "solutions" to classroom practitioners; that "solutions" developed centrally have universal effectiveness; and that there is a "passive" consumer at the end of the delivery system' (Robottom 1987).

Sadly, this description sounds a very long way removed from the graphic and moving accounts of developing environmental awareness and concern through such experiences as having sermons from beech trees and opening one's mind to greater things outside ourselves that were mentioned at the outset of this chapter.

Finally, there are significant practical problems associated with the incorporation of a 'peripheral' (non-statutory) element into an already crowded curriculum. The National Foundation for Educational Research in England undertook a number of specific research projects in environmental education whose results included the following:

- Only 7% of schools had produced a specific environmental education policy. 42% had no environmental education of any sort.
- Less than 25% of schools has a co-ordinated, cross-curricular approach across many subjects. Geography and science curricula were used as the main vehicles for environmental education.
- The main constraints were identified as lack of timetable time (because of the need to meet statutory requirements); lack of resources; lack of staff expertise and lack of staff motivation.

(Tomlins and Froud 1994)

It seems very clear that all of these factors have in common the tendency to alienate the spiritual dimensions of learning and human development from programmes of environmental education.

EDUCATION FOR ENVIRONMENTAL INSIGHT

So let us now return to this chapter's main theme; and to the rather disjointed picture of factors affecting the development of environmental awareness that arises from our empirical research – a scene wherein for many individuals, ideas and personal experiences in the spiritual and religious domains have had as great or even greater impact upon their thinking in relation to the environment as formal education; where formal environmental education is indeed playing an important role, though for various reasons it is not as successful as it should be; and where there is often a false separation of these two spheres of influence upon people's lives. Fundamental questions remain. Are there ways in which these two strands of experience can be more closely integrated? Can there be a revaluation of the role of spiritual development? Can teaching and learning about the environment encompass a much broader vision of environmental issues? I conclude with some tentative suggestions for ways forward, if only to encourage reflection and debate.

Solutions to the questions raised above can obviously be addressed on a variety of scales. At the 'macro', radical level of analysis, one questions the contemporary goals and purpose of schooling; and teachers' and curriculum ideologies and practice. Should we be moving towards an 'ecological paradigm' for education in which education is shaped by an ecological world view? Such a paradigm shift is towards what Arne Naess (1973) describes as 'deep ecology', which cultivates a state of being that sustains the deepest possible identification of oneself with one's environments or an 'ecological consciousness' (Devall and Sessions 1985). This sense of identification with nature is so profound that individuals see no separation of themselves from nature; both are part of 'being'. In such a culture environmental education, as we define it, would be unnecessary, since ecologically desirable actions would naturally 'fall out' of this state of being rather than follow from some moral or intellectual imperative (Fox 1986).

Such thinking has led a number of writers to develop the concept of an ecological paradigm for education; effecting a shift away from the epistemological paradigm influenced by society's dominant form of theorizing, namely, positive empirical science, to a paradigm based on ecological theories of perception. Gough (1987) comments that empiricist theories of perception suggest that human perceptual systems provide no reliable knowledge of our world, which can only come through analytical abstraction and logical inference. Knowledge is socially structured, largely theoretical and technical. Ecological theories of perception, on the other hand, suggest that human perceptual systems have evolved to detect informational structure of environments and become more sensitive through practice in perception. Knowledge is individually structured, practical and personal. According to Emery (1981), the object of learning in an epistemological paradigm is the transmission of existing knowledge, and abstraction of generic concepts. In an ecological paradigm it is the perception of invariants, the discovery of serial concepts, and the discovery of universals in particulars in learners' environments. Proponents of the need to promote such a paradigm change identify small but significant steps that could be taken to improve practice, even in the short term:

> First, we can try to trust our personal subjective experiences rather than defer habitually to the entrenched status of accumulated propositional knowledge. Second, we can try to educate our own senses so that we become better at searching out the characteristics of the personal, social, and physical environments in which we conduct our educational practices. Above all, we can try to see such searching – like all learning – as a relationship . . . we should have more faith in what we can learn *with* environments.
>
> (Gough 1987: 64)

Thus we move in the direction of strengthening of beliefs in the value of inner, subjective experiences; of closer integration between spiritual ideas and the practice of environmental education.

Rethinking education – or even just environmental education – according to an ecological world view, may well sound hopelessly long-term and idealist. So what might realistically be achieved according to a less radical level of analysis? Even without a major paradigm shift, surely it is possible for education policy and practice to incorporate more of what I will term 'environmental insight'; for institutions to move towards a more humane, ecological and person-centred vision both of society and the curriculum they teach. We have the inspiration, the goals and the guidance in some of the leading global documents of the recent past; for example, *Our Common Future* reminds us that sustainable development requires changes in values and attitudes towards environment and development:

> The world's religions could help provide direction and motivation in forming new values that would stress individual and joint responsibility towards the environment and towards nurturing harmony between humanity and environment. . . . Education should therefore provide comprehensive knowledge, encompassing and cutting across the social and natural sciences and the humanities, thus providing insights on the interaction between natural and human resources, between development and environment.
> (World Commission on Environment and Development 1987: 111, 113)

Caring for the Earth: A Strategy for Sustainable Living underlines the importance of values and world ethic:

> Environmental education deals with values. Many school systems regard this as dangerous ground and many teachers (particularly in the natural sciences) are not trained to teach values . . . it is vital that the schools teach the right skills for sustainable living . . . it is equally important that what the school does reinforces what it teaches. . . .
> Every society is likely to have special symbols, stories, sacred places and other cultural features that can support the world ethic for living sustainably as well as its own cultural needs. These should be identified, so that educational programmes can be tailored to the culture and environment of the society that they serve.
> (IUCN, UNEP, WWF 1991: 55, 53)

Agenda 21 of the UNCED talks of reorienting education towards sustainable development:

> To be effective, environment and development education should deal with the dynamics of both the physical, biological and socio-economic environment and human (which may include spiritual) development, should be integrated in all disciplines, and should employ formal and non-formal methods and effective means of communication.
> (Quarrie 1992: 22)

Regrettably, such inspiration and guidance is rarely translated into practice; yet surely a redefinition of the place of environmental understanding
in a traditional curriculum to include its planned and structured incorporation into the arts and humanities as much as into science would go a
long way towards the successful permeation of 'environmental insight'
into day to day educational routine. Yes, of course, there is every place
for the acquisition by pupils of a scientific, ecological perspective – the
learning of a range of systematic knowledge from the traditional disciplines of biology, physics, chemistry, technology, geography and geology.
In the traditional model of environmental education, this is described as
education *about* the environment, that is, the transmission of knowledge
that may explicate aspects of the environment or provide conceptual
capacity to do this. Environmental education incorporates this and far
more besides. The gaining of true environmental insight entails a far wider
perspective on issues affecting our planet with substantial emphasis on
the human, including spiritual, dimensions of interaction with the environment. Dramatic increases in the human population and its activities have
changed the environment more rapidly in the late nineteenth and twentieth centuries than at any other time in our history. Challenging issues
for serious study in the curriculum for history could well include the
ways in which major upheavals such as the Industrial Revolution have
transformed the environments people live in and the impacts such transformation has had on all aspects of their lives. Other aspects of the history
curriculum could focus on the history of our treatment of the environment
as something of interest in its own right – how and why did things in the
environment we take for granted, such as gardens and nature reserves,
come about? What does their emergence tell us about the societies and
attitudes to nature of the times? Then, going beyond a study of our
treatment of and attitudes to the environment for its own sake – what
about the impact such attitudes and behaviours have had on political
ideas and actions? In religious education, studies could look not only at
various religious attitudes *towards* the environment, but at the extent to
which religious beliefs are shaped *by* the environment. How interesting
it would be, for example, to consider why the monotheistic religions of the
world originated mainly in harsh, 'unfriendly' climates, while polytheistic
religions developed in warmer climes. The art curriculum could contain
an element of focus on how various art forms have treated and interpreted
the environment; conceptions of the way landscapes have been painted,
and so on – and another on the impact of art *on* the environment – what
people have done to landscapes in order to transform them aesthetically.
Music lessons lend themselves to the obvious questions of the extent to
which landscape and other aspects of the environment affect the composition of music, and how such things are portrayed *through* music and
song. Similar questions could be considered in the field of literature,
alongside a detailed look at such things as how poetry enables reflection

on beliefs and the human situation; 'romantic' traditions in literature, why some countries have a far stronger tradition of intellectual concern for nature than others, and so on.

It is not being suggested that activities such as those mentioned above as examples are necessarily left undone – rather; that in the majority of situations they are left to chance – to the slant of a particular syllabus, the choice of a topic, the enthusiasm of a certain teacher. As long as environmental education remains peripheral to a core curriculum, its content specified as 'belonging to' the realms of science and geography, then its spiritual and ethical dimensions will often-times be unplanned and spasmodically addressed.

Broadening the base of environmental education into this structured permeation of environmental insight across the curriculum should successfully dispel the myth that environmental educators are concerned solely with teaching about the environment (as biologists are concerned with teaching about life in the biosphere, chemists with teaching about chemicals, musicians with teaching about music, and so on). Environmental education is as much about interactions between human life and the environment – about insights, intuitions, attitudes and emotions connected with the desire to overcome and prevent detrimental environmental impacts in the future – as it is about measuring sea-level changes and holes in the ozone layer.

Widening perspectives on environmental education in this way clearly has implications for curriculum design and policy at the highest levels. The National Curriculum document on spiritual development mentioned at the outset of this chapter (NCC 1993), identifies three areas of school life in which opportunities arise for spiritual and moral development: the ethos of the school, all subjects of the curriculum and collective worship. 'Moral issues will arise, for example, in science (issues of life and death), geography (environmental issues), and history (development of tolerance).' A clear impression is given here that environmental matters are compartmentalized, 'attached' to a single-subject discipline; when in reality the development of environmental understanding and of spiritual ideas and feelings can go hand in hand, supporting and enhancing each other in a variety of learning situations across the whole spectrum of the curriculum.

To draw this account to a close, I turn from discussion of what might be desirable to an exciting example of what *is*. The *Core Curriculum for Primary, Secondary and Adult Education in Norway* (Hagness 1994) constitutes a legally binding foundation for the development of separate curricula and subject syllabuses at the different levels of education – the common core for the Norwegian education system. According to this policy, education of 'the integrated human being ... shall inspire an integrated development of the skills and qualities that allow one to behave morally, to create and to act, and to work together and in harmony with nature. Education shall contribute to building character which will give the individual the strength to take responsibility for his or her own life, to make a commitment to society, and to care for the environment.' Sections within the document, thus constituting the curriculum foundation, are 'the spiritual human being', 'the creative human being', 'the working human being', 'the liberally educated human being', 'the social human being' and 'the environmentally aware human being'. Environmental understanding and awareness is there at the heart of the national core curriculum. Furthermore, within the section on the environmentally aware human beings, descriptions within its subheadings – Natural Science, Ecology and Ethics; Human Beings, the Environment and Conflicts of Interest; and The Joy of Nature – make it quite clear that this curriculum anticipates the crucial role of spiritual ideas and experiences in developing environmental awareness:

> Education must also kindle a sense of joy in physical activity and nature's grandeur, of living in a beautiful country, in the lines of a landscape, in the changing seasons. It should awaken a sense of awe towards the unexplainable, induce pleasures in outdoor life and nourish the urge to wander off the beaten track and into uncharted terrain; to use body and senses to discover new places and to explore the world. Outdoor life touches us in body, mind and soul. Education must corroborate the connection between understanding nature and experiencing nature.
>
> (Hagness 1994: 38)

A curriculum based on statements such as this, with the structured inclusion of environmentally focused work across its whole spectrum of activity would surely serve as an ideal forum for the nurturing and development of personal experiences in the natural world – experiences which, according to a majority of research subjects, are deeply significant in the development of environmental awareness and concern.

12 Spirit of middle earth
Practical thinking for an instrumental age

Richard Smith

And hence this Tale . . .
                    . . . by the gentle agency
Of natural objects led me on to feel
For passions that were not my own, and think
At random and imperfectly indeed
On man; the heart of man and human life.

Wordsworth, 'Michael': lines 27–33

Middle earth: The earth as placed between heaven and hell. Now only
arch., occas. applied to the real world as dist. from fairy-land.

(Shorter Oxford English Dictionary)

Two particular pitfalls lurk for those who write these days about spiritual
and environmental education and their interconnections, as I attempt to
do here. One is the descent into the dispiriting jargon of modern education. I illustrate and discuss this in Part I below. The other is the
portentousness which makes human spirituality and our experience of
the natural world into a rapt mystery, as if nothing less will do than
dumbstruck contemplation of mountains, sunsets and starry heavens.
From this springs talk of 'awe and wonder', a phrase all too familiar from
recent writings.

At the root of the jargon, I shall argue in Part II, is our predominantly
instrumental world view, while the 'awe and wonder' school of thought is
a version of romanticism that reacts, and overreacts, against that. There
is, of course, nothing original in seeing our salvation, environmental and
spiritual, as lying in a middle way between instrumentalism and romanticism. Yet the extent to which our relationship with the environment in
that middle way can be seen as a properly *spiritual* one is perhaps less
obvious and worth trying to explain. In Part III I try to show that in the
middle way we can find a recognizable, uninflated kind of spirituality that
will be nourished by a well-founded relationship with the natural world
of fields, rivers and animals, particularly in the context of human *work*,
and that will in turn serve our environmental concerns well. Part IV gives
an extended picture of such a relationship, and connects it with the idea

of spiritual 'attentiveness'. Lastly, in Part V I suggest certain conclusions for the way that we think of language and use it.

Language is a major theme of this chapter: both pitfalls referred to above exhibit language in collapse. In the first language is used so crassly, with so little sensitivity to the nuances of human experience, that those subtle dimensions of it which we call 'spiritual' can only, we may think, be damaged and rendered still less accessible. In the second – and here again we can see the reaction to the first – there is a tendency to see spirituality as beginning where language ends. This, too, is to give up on language, as though in our language-saturated world we could do without it just when things get really important.

Part I goes into linguistic detail to a degree that some may think excessive. If my argument in this chapter is at all persuasive, however, the reader may come to share my view that such insouciance is one more symptom of our environmental and spiritual crisis.

I

It is easy enough to agree with the idea that education should be as concerned with children's spiritual development as with the development of their intellectual, moral and other capacities. Indeed, the Education Reform Act of 1988 specifically requires the National Curriculum of England and Wales to promote children's 'spiritual, moral, cultural, mental and physical development'. The suspicion that lip-service is being paid here, however, is raised by the fact that the National Curriculum is essentially one focused on school subjects, traditionally conceived as English, mathematics, science, and so on. Together with environmental education, spiritual education risks being swept to and even beyond the furthest margins of formal schooling. Despite the legal status of spiritual education to which I have already referred, and although environmental education was recognized as an important 'cross-curricular theme' in the National Curriculum alongside such other themes as health education and education for citizenship, neither figures prominently in recent documentation. For example, a recent *Guide to the National Curriculum* (SCAA/ACAC 1996: 3, 63) makes no mention whatsoever of environmental education and the other cross-curricular themes, and the only references to spirituality occur in an acknowledgement of the requirements of the Education Reform Act and in the section on religious education. The one connection made between that and the spiritual life, in three pages, is the indication that religious education aims, among other things, to help pupils 'enhance their spiritual and moral development'.

Despite the wider neglect of issues of spiritual development there are from time to time flurries of concern. Often these appear to be prompted by events in the news, such as the murder of a headteacher by local youths, or the violence that we periodically remember children can inflict on each other. In what follows I discuss extracts and themes from two recent documents which seem to have originated in such outbreaks of concern. The first extract comes from a National Curriculum Council 'Discussion Paper', *Spiritual and Moral Development.*

Spiritual development in an educational context

Spiritual development is an important element of a child's education and fundamental to other areas of learning. Without curiosity, without the inclination to question, and without the exercise of imagination, insight and intuition, young people would lack the motivation to learn, and their intellectual development would be impaired. Deprived of self-understanding and, potentially of the ability to understand others, they may experience difficulty in co-existing with neighbours and colleagues to the detriment of their social development. Were they not able to be moved by feelings of awe and wonder at the beauty of the world we live in, or the power of artists, musicians and writers to manipulate space, sound and language, they would live in an inner spiritual and cultural desert.

(NCC 1993: 3; punctuation and syntax as original)

A chief point of interest in this passage is the way that an elevated *lexis* (spiritual, imagination, awe, wonder, beauty, space) finds itself at the mercy of a syntax which, while aiming at noble effect (as if this would somehow guarantee that *spiritual* matters were at issue), staggers clumsily through the paragraph. At the same time the content, which we expect to emphasize the importance of the spiritual life, frequently descends to bathos. The first two sentences promise to celebrate spirituality, but show a strong tendency, as does the somewhat opaque third sentence, to assert that spiritual development is essentially valuable insofar as it contributes to other kinds of development: intellectual, perhaps academic, and social development. (We should have been warned by the heading, with its near-oxymoronic collocation of 'spiritual development' and 'educational context'. What would we expect of '*Karma* in the classroom' or '*Agape* across the curriculum'?)

All the sentences except the first bring their subordinate clauses forward to the beginning, a technique of high prose style designed to maximize the impact of the deferred main clause for which we have to wait in expectation. This device is in the event let down badly by the tautology of the ending of the third sentence – if they had difficulty co-existing with others would this not be what poor social development *consisted in*? – and, in the second half of the fourth, by the jarring note struck by 'manipulate space, sound and language' with its suggestion that here we have a list of *skills* and that artists, musicians and writers proceed with the degree of detachment and deliberateness implicit in 'manipulate'.

Most uncomfortable of all is the last clause, the consummation for which we have been waiting. Children might live in a spiritual and cultural desert, but not an inner one. An inner one would be found in them. The high tone the passage seeks is further undermined by infelicity of punctuation and the use of 'may' before 'experience difficulty'. Here, in conformity with 'would lack', 'would be impaired', 'would live', we expect not 'may' but 'might'. 'Might', I suspect, has lost popularity because it has a slightly nasal and ungenteel twang. Accordingly we now hear such sentences as the football commentator's 'had the defender not timed that tackle right Shearer may have scored' when there is, in fact, no doubt that the ball is safely down the other end of the pitch.

There are four features of this passage that I am concerned to emphasize. The first is the prevailing *instrumentalism*. Spiritual development is desirable less for its own sake than for the sake of other kinds of development – intellectual and social – which are known in advance and probably specifiable in a degree of detail. Writers and artists are technicians, heaving space, sound and language around like steel girders or pre-stressed concrete. They make these materials do their will for purposes which they can predetermine: this is part of the meaning of 'manipulate'. The second feature is the faint suggestion, which we shall find comes across more strongly in other extracts examined below, of a kind of educational jargon operating as ultimate justification: as if talk of intellectual and social development (possibly to be articulated in terms of progression through Key Stages and the promotion of self-esteem respectively) formed an uncontroversial foundation for discussion of other, more elusive matters such as spirituality. The third feature, perhaps an unconscious compensation for this crude reductionism, is a characteristic *cant*, combining a style both genteel and pompous with a tendency to moralize as well as a reliance on grandiloquence to get the writer out of a tight spot. And fourthly there is a naivety about language: a belief that you can write about spirit, establish a bureaucracy of spiritual education with attendant nomenclature, in the same way as you can with, say, mathematics or equal opportunities.

I do not believe that these are mere points of detail, minor blemishes at most that can be overlooked in the broad welcome we should extend to a renewal of interest in things spiritual. We see here rather some of the forces that operate to diminish both our understanding and our very experience of the life of the spirit.

Some confirmation that these features really are deep, underlying tendencies in education comes from the fact that they can be found in a second document, *Education for Adult Life: The Spiritual and Moral Development of Young People* (SCAA 1996). It is notable that this 'summarises ideas expressed' at a conference of the same name, with the result that the author or authors are able to shift between reporting delegates' views and authorial assertion in a manner that obscures the subjectivity of the latter. As illustration of this, and of the same rampant instrumentalism we have already noted, consider the following passage:

Spiritual development and learning

Some delegates regarded all learning as spiritual activity. A spiritual sense can be seen as a prerequisite for learning, since it is the human spirit that motivates us to reach beyond ourselves and existing knowledge, to search for explanations of existence. The human spirit engaged in a search for truth could be a definition of education [*sic*], challenging young people to explore and develop their own spirituality and helping them in their own search for truth.

If the association between spirituality, moral development and learning is accepted, the importance of spiritual development at school is self-evident.

(SCAA 1996: 6)

Or consider this remarkable extract, bulleted as *Materialism and greed*:

To some delegates, a money-centred value system promotes selfishness and the desire for power. There was a fear that technological and economic advances have marginalised spiritual values and created a society in which individual success is evaluated only by wealth and status. The paramount values are primarily selfish, and centre on acquiring money and material items. There was widespread agreement that such values could undermine the well-being of society and its supporting economy.

(SCAA 1996: 9)

And its supporting economy! As if the value of the life of the spirit came down to this, that its marginalization threatens the economy!

II

In the section above I have criticized the predominant 'instrumentalism' of writing about spirituality. We have become familiar enough with critiques of instrumental or 'technical' reason. Its spectacular success in devising means for predetermined ends and so helping us to achieve technological control over the world has led us to think of *all* rational action as based on means–end calculation. This brings about the marginalization of the ethical dimension of our lives, since values are held to inhere only in ends while the means that lead to them are neutral. In the means–end thinking that constitutes instrumental or technical reason the very possibility of deliberation about fundamental questions of ends is relegated to the status of personal preference, of 'values and attitudes', and the really important business appears to be that of working out how to achieve ends that have become largely taken for granted.

With the domination of instrumental patterns of thought and action comes the loss of a sense of meaning. Instrumentalism is responsible for this by foregrounding more superficial forms of meaning over profounder ones, so that we find the latter harder to perceive (see Taylor 1989: 500). What has happened is that 'by dissolving traditional communities or driving out earlier, less instrumental ways of living with nature, [it] has destroyed the matrices in which meaning could formerly flourish' (ibid.). This brings about that 'disenchantment' in which 'the world, from being a locus of "magic", or the sacred . . . comes simply to be seen as a neutral domain of potential means to our purposes' (ibid.).

This 'disenchantment' is partly the product of the breakdown between us and the natural world, and partly the cause of it. 'The life of instrumental reason lacks the force, the depth, the vibrancy, the joy which comes from being connected to the élan of nature. But there is worse. It doesn't just lack this. The instrumental stance towards nature constitutes a bar to our ever attaining it' (ibid.: 383). We come to see ourselves as deeply and irrevocably separated from nature, and so as prevented from opening ourselves up to it. We find it difficult to recover a sense of the demands that the natural world, and our relationship with the animals, make on us. This, in turn, feeds our tendency to objectify nature, to see it as a neutral realm to be used, operated on, dominated.

So there arise those modern attempts to repair the divorce between ourselves and nature that can be characterized as forms of neo-Romanticism. Certain forms of feminism, concerns for self-fulfilment and psychic wholeness, what Charles Taylor calls the 'human potential' movement and Christopher Lasch the 'culture of Narcissus' all have this in common, that they reject instrumental rationality and the technological 'Prometheanism' that elevates Man decisively and damagingly over nature. At the environmentalist end of these tendencies versions of 'deep ecology' and the well-known 'Gaia hypothesis' of James Lovelock attempt to solve problems in the relationship of humankind to nature by the sweeping device of making the bearer of rights the planet as a whole. In this way troublesome questions of the rights of people versus those of animals, trees or mountains cannot arise.

It is common to find these neo-Romantic yearnings linked with perspectives on nature which it is claimed have been held by native peoples. The Hopi and Australian aboriginals are frequently cited. The following passage from Bowers (1993: 143) can stand for many similar ones:

> [For native peoples] Learning how to live in a habitat . . . involves learning from elders (survivors), generations no longer present, plants, animals, soils, weather patterns, and all other elements of the habitat. Knowledge, context, continuity, and practice seem to be intertwined

and holistic. . . . Unlike the emancipatory orientation of critical reflection, which incorporates Western assumptions about knowing being based on an individual perspective and a distancing/objectifying relationship . . . [in traditional forms of knowing] the person is not viewed as the primary repository of knowledge.

Bowers draws on the work of Gregory Bateson to argue that such an ecological perspective leads us to see that the individual mind is only a subsystem of that larger one which has been called, for example, God or the Great Spirit. 'For traditional peoples this awareness of the interconnected planetary ecology represents a form of spiritual knowledge that provides the basis for living a moral life; that is, one that is in harmony with the patterns of sustainable ecological order' (ibid.: 144). Nothing could be clearer: the environmental attitudes characteristic of (many? all?) native peoples *are*, in this view, a major dimension of the spirituality whose loss we now keenly feel. What will save the planet will also save our souls. And a further, optional, twist connects the 'strong oral orientation' of native peoples with a deep sense of community and more contextualized ways of thinking. Bowers insists that he is 'not suggesting that ecological sustainability will require the abandonment of literary-based discourse', but he is happy to suggest that 'if teachers give more emphasis to involving students in the patterns associated with oral culture, they may be reinforcing micropatterns that will be consistent with the macropatterns of an ecologically sustainable culture' (ibid.: 200). Thus where language – especially that of the Western, instrumentalist tradition – has failed us by reproducing objectifying, exploitative relationships, the silent stupefaction of 'awe and wonder' may well, to the neo-Romantic, seem preferable.

Nothing in what I write is intended to denigrate the world-views of native or indigenous peoples, though it seems legitimate to ponder how far the ecological sustainability of their cultures follows from the simple fact that they did not develop the technologies which have proved such mixed blessings to the developed world. There is a danger, too (and one which Bowers himself clearly recognizes), of imagining that we can do any good by trying to graft elements of native traditions on to our very different culture: a Hopi assembly in the primary school, say, or a short project on the Dreamtime. This, in the terms of my title, is fairy-land. For better or worse we have moved on, and there is no road back. Where are we denizens of middle earth to find our spiritual and ecological salvation?

III

In our reaction against instrumental rationality and in our search for meaning and the lost sense of spirituality perhaps we do not need to go to the extremes of neo-Romanticism. As a number of writers have reminded us recently, there is a great deal to be gained from a recovery of the idea of *practical reason* (also called, by some, practical knowledge or practical judgement), the Aristotelian *phronesis*. The hallmarks of practical reason are flexibility and attentiveness to the details of the particular case (Aristotle calls this attentiveness *aisthesis*, sometimes translated as 'perception' or 'situational appreciation'). It is coloured by sensitivity and, crucially, *attunement* towards its subject-material, rather than the attempt to exercise mastery or control over it. The craftsman, for instance, has a certain 'feel' for the wood or stone he is working, and knows that if forced it will split or shatter. Instrumental or technical reason produces goods which are specified by criteria that lie outside the process of making. The car that comes off the assembly-line is determined by considerations of what can be sold to the customer; the manufacturer is unlikely to be moved by the thought that a different way of going about the process will help to keep alive certain craft-skills among the workforce. In practical reason, on the other hand, we seek the good that we attempt to realize *through* the action and not as a separate and independently identifiable aim. Christopher Lasch (1984: 253) puts the point memorably: 'the choice of means has to be governed by their conformity to standards of excellence designed to extend human capacities for self-understanding and self-mastery'.

For Aristotle the paradigm case here is politics. The good nation-state is one where we find fair and honest dealings among its statesmen and its citizens: it is not simply one which successfully pursues such ends as a low inflation-rate and the defence of its territorial boundaries. To construct a modern example, we can say that the good farmer does not cultivate his land with an eye only to its short- or even its medium-term yields, irrespective of the harm done to wildlife or the damage done by depletion of aquifers, since he has a sense of the standards conducive to (in Lasch's phrase) 'self-understanding and self-mastery'. These standards include a keen awareness of the dangers of greed and an exploitative attitude. It is not just that the farmer knows how much greed on his part will damage the natural world. Rather he is aware that greed is a snare for a man who wants to live in harmony with himself. In the case of a farmer especially, greed will pull against those traditional agrarian virtues – prudence, thrift, a certain steadiness – in terms of which he understands himself and his place in the world. Greed will harm the land but it will harm, too, we might say, his soul.

In this way practical reason is irreducibly ethical, and its ethical quality is of a rich and complex kind, involving a continuous, if not always fully conscious, testing of one's action against the internal goods of an activity. Yet there is more than an *ethical* dimension to practical reason, or certainly more than what the late twentieth century has come to think of as the ethical (which is typically a 'bolt-on extra' to be reached for at times of moral panic). The sensitivity or attunement towards its subject-matter characteristic of this form of rationality seems to take us back to the pre-modern era where our understanding of the order of things is bound up with our understanding of our own place in it, since we are part of that order (see Taylor 1983: 142). 'And we cannot understand the order and our place in it without loving it, without seeing its goodness, which is what I want to call being in attunement with it' (ibid.).

It is important to recognize that this perspective does not involve returning to a world of medieval mysticism where (for example) the oneness of humankind with the universal order extends to the authority of planetary influence over our affairs. The notion of attunement is entirely at home in, for example, those conceptions of human work which allow us to develop a 'feel' for our materials and, in refining our skills, to exercise and nourish the mundane and unpretentious virtues – virtues of middle earth indeed – that belong there. The carpenter knows his wood and has respect for its qualities. The experienced cook employs her knives, pans and other equipment so to speak as an extension of herself and comes to know quite instinctively which flavours complement which. Like all craftsmen the carpenter and the cook must be patient, meth-odical, sometimes extemporizing ingeniously. This, of course, is to talk of those in a position to practise their craft with a sufficient degree of autonomy and creativity. It is precisely because the fulfilling *work* in which these crafts can be practised has so widely been replaced by mech-anized, repetitive and alienated *labour* (or 'drudgery') that we both forget how readily available these possibilities of attunement have been until recently and at the same time incoherently romanticize the craft traditions in which they are found. (One sign of this is our tendency undiscriminat-ingly to call all manner of things, from the lowliest knacks to the profoundest human qualities, 'skills': teeth-cleaning skills, parenting skills, love-making skills. . . .) So thoroughly has modernism broken the connec-tion between rationality and attunement, and left us more than *ethically* adrift.

IV

The opening lines of Wordsworth's poem 'Michael' speak of a 'hidden valley' among the mountains: an 'utter solitude' which the poet would not have mentioned but for a 'heap of unhewn stones' that have the story of the shepherd, Michael, attached to them. The story is just one of many stories of shepherds: 'men / Whom I already lov'd, not verily / For their own sakes, but for the fields and hills / Where was their occupation and abode' (lines 23–6). Neither nature nor mankind is interesting to the poet on its own, but only insofar as nature bears the mark of man and man in turn is attuned to nature.

Nor is it only the Romantic poet who can achieve this attunement.

Wordsworth tells us that it would be a 'gross mistake' to suppose 'That the green Valleys, and the Streams and Rocks / Were things indifferent to the Shepherd's thoughts' (lines 63–4). The fields and hills 'Which were his living Being... /... were to him / A pleasurable feeling of blind love, / The pleasure which there is in life itself' (lines 75–9). Michael is a man of practical reason or wisdom, with the virtues of his calling:

> ... his mind was keen
> Intense and frugal, apt for all affairs,
> And in his Shepherd's calling he was prompt
> And watchful more than ordinary men.
>
> (lines 44–7)

He has the sensitivity and responsiveness to find practical meaning in 'all winds' (line 48), knowing which mean danger to his flock and strenuous work for him. He, his wife and son Luke are proverbial in their community 'For endless industry' (line 97): here perhaps is an echo of the Aristotelian idea that people of practical reason attend to the internal goods of their practice and do not constantly think of the external aims or ends.

The Romanticism that has descended from Wordsworth and his contemporaries has foregrounded the works of nature, and the world of rustic people, as especially apt for the generation of spiritual insight. Wordsworth himself (for example, in the Preface to the *Lyrical Ballads*) sometimes writes in a way that lends itself to this interpretation. A further twist takes the spiritual out of the realm of the ordinary and quotidian towards intense and quasi-religious experience. Thus Coleridge famously claimed (*Biographia Literaria*, ch. 14) that Wordsworth wanted 'to excite a feeling analogous to the supernatural, by awakening the mind's attention from the lethargy of custom, and by directing it to the loveliness and the wonders of the world before us'. This is the road that leads to 'awe and wonder'. Yet we can equally understand 'Michael' as a poem that points to the possibility of finding spiritual significance in the world of ordinary, mundane things and in human work, properly understood. The merging of mind with nature regularly asserted to be the touchstone of Wordsworthian Romanticism is neither possible nor desirable where man has rather to maintain a degree of separation from the natural world in order to do his work in it, albeit by the standards of practical rather than instrumental reason.

In the poem the shepherd Michael finds himself financially burdened and must let his son go away to work for a kinsman. On the day before parting the old man takes Luke to where he has gathered stones to build a sheep-fold, and asks him to lay the cornerstone as a covenant between them:

let this Sheep-fold be
Thy anchor and thy shield; amid all fear
And all temptation, let it be to thee
An emblem of the life thy Fathers liv'd.

(lines 417–20)

But Luke, away in the 'dissolute city' (line 453), goes to the bad, and though Michael still visits the site of the sheep-fold and works at building it 'from time to time' (lines 469 and 479) it is unfinished at his death. Torn between the love of his son and love of his inherited land Michael has attempted to join the two by an almost magical device, the sheep-fold that is to be both a work of practical utility and, for his son, something between a symbol and a charm.

Thus even practical reason can fail us spiritually, especially if we strain too hard to invest it with spiritual significance.

Among the rocks
He went, and still look'd up upon the sun,
And listen'd to the wind; and as before
Perform'd all kinds of labour for his Sheep,
And for the land his small inheritance.

(lines 464–8)

But if he listens to the wind he does not appear to hear anything there. The meaning is gone.

The world of nature may be a particularly fitting place to exercise our practical reason, and it is there, of course, that our environmental and spiritual concerns may happily come together. But it is important to emphasize again that the 'attunement' of practical reason may be found in a great range of human activities. The gardener gets a 'feel' for the ideal depth and consistency of earth to cover the seeds; the dressmaker knows as it were through her fingers how tight the stitching should be, with this particular thread for this particular material. The carpenter, mentioned above, can stand as representative for 'any careful attentive self-forgetting work or craft, including housework, and all kinds of nameless "unskilled" fixings or cleanings or arrangings which may be done well or badly' (Murdoch 1992: 180). This can be true even in the world of technology, where we are not obliged, nor always best advised, to use technical or instrumental reason only. As Pirsig's remarkable novel *Zen and the Art of Motorcycle Maintenance* shows us vividly, practical reason and its attunement to its material may be found in our relationship with the most unromantic of artefacts. The craftsman, says the novel's narrator, is:

absorbed and attentive to what he's doing even though he doesn't deliberately contrive this. His motions and the machine are in a kind of harmony... the nature of the material at hand determines his thoughts and motions, which simultaneously change the nature of the material at hand.

(Pirsig 1975: 167)

This is as true for our relationship with the motorcycle of the title as for wood, seedlings and fabric.

The idea of attention or 'attentiveness' seems important here. Properly to attend, to things or to people, means putting aside the demands of the insistent, selfish ego. Attending carefully to a particular person we can come to see him as refreshing rather than simply naïve, as someone with a certain integrity rather than just a crank. Literature often requires us to bring a just attention to bear: on Fanny Price in Jane Austen's *Mansfield Park*, to take a celebrated instance, who can appear either as a prig or as a woman of quiet principle; or on Stevens, the butler in Ishiguro's *The Remains of the Day*, whose loyalty to his master can seem by turns pathologically blind or, in an old-fashioned sort of way, rather admirable. Iris Murdoch has followed Simone Weil in writing extensively about the significance of attention. People or situations can be seen in different lights if we exercise patience, determine to see them justly, and refuse the consolations of fantasy. Even the faulty computer can be *interesting* rather than an infuriating nuisance if approached in the right, as we say, spirit. The 'small contingent details of ordinary life and the natural world' (Murdoch 1992: 244) may be loved and respected, as Zen Buddhism teaches. The transformation in the energy and vision of the person who attends justly, who 'looks in the right light', seems more than a (merely) ethical matter. This may not quite be the kind of spirituality, derived from the high Christian tradition, which is now commonly thought to be in decline, particularly in the world of education, and in need of reinstating. But it does appear to be a kind suitable for our lives here on middle earth, on a fragile and threatened planet.

V

The instrumental or technical forms of reason which are our dominating forms of thought, and which I argued in the first section of this chapter can be found rampaging through recent writing about spiritual education, can also be seen in many of the ways in which we think about language itself. Just as we think calculatively about the natural world, and how to turn it to our purposes, so we tend to think of language as a means of order and control. This is especially true in education where a major tendency in writing about language in education from the early 1970s onwards has been to regard language as essentially a device aimed at 'detection and conquest of an external world' (Cassirer, quoted with approval by Britton 1970). The influential Bullock Report of 1975, *A Language for Life*, prefixes a chapter on 'Language and Learning' with a quotation from Gusdorf: 'Man interposes a network of words between the world and himself, and thereby becomes the master of the world'. The tendency has if anything become even more pronounced in recent years: one sign among many is the regular insistence on the 'effective' use of language (six occurrences in the first seven pages of the 1988 Kingman Report on language, for example). 'Effectiveness' threatens to establish instrumental, means–end reasoning as the *only* way to think in the 'real world'.

This way of conceiving language is extremely deep rooted. It is there in the common supposition that language is primarily or only a means of communication, as if we had antecedent thoughts and then struggled for the best – most effective – words to put them in. It is there in the apparently commonsensical notion that language is 'designative', picking out aspects of an antecedently individuated world: a notion challenged by the Sapir-Whorf hypothesis that language creates our experience of the world rather than reflecting it.

If the instrumentalism that can plausibly be said to be at the roots of our environmental crisis and the collapse of spirituality infects our very understanding and use of language, then we are in very deep trouble, for all our efforts to articulate our problems and find solutions to them risk being corrupted by the same virus and spreading it further. This, I suggested in the first section of this chapter, is what has happened with recent writing about the education of spirituality. Faced with this, we may very well be inclined to agree with John White (1995: 16), who would advocate 'an absolute embargo on the use of the terms "spirituality" or "spiritual development" in all official documents on education'. Such full-blown secularism, however, concedes too much ground to instrumentalism. The task we are faced with is one of recovering ground that has been lost rather than giving up more and more, and if language itself has been the central and crucial loss then this is where we have to start.

We might begin with literature. Poetry, or the novel, was once seen as something you immersed yourself in, surrendered yourself to. In other times texts – the Homeric epics, for example – were *authoritative*. Shakespeare, it was held, was a writer whose contemporaries we were struggling to become. Poetry was often learned by heart: like the language of the Bible it entered the bloodstream. John Keats' poetry is full of Shakespearean echoes, less as a matter of conscious artistry than because he was awash with the language and images of Shakespeare's plays and sonnets. In this, now superficially old-fashioned, way of thinking of literature there are no doubt dangers, especially of that uncritical excess of respect which now culminates in notions of 'our literary heritage'. Yet its replacement by ideas in which the reader, or the literary or social critic, seems to be accorded more authority than the text constitutes the loss of a major dimension of a non-instrumental philosophy of being in the world.

In that philosophy it becomes impossible to distinguish language and literature (that we now make this distinction so easily is another sign of the trouble we are in). So far from being a tool at man's disposal to use instrumentally as he wills, rather 'language speaks man', in Heidegger's phrase, being that which reveals the hiddenness of the world if we will only open ourselves to it. 'Man acts as though he were the shaper and master of language, while in fact language remains the master of man' (Heidegger 1971: 215). Language is less like an instrument than a woodland path, a *Holzweg*, for us to follow wherever it goes.

It is possible to glimpse here how a non-instrumental stance to language might prepare us for a stance towards the environment which similarly allows for other stances than the instrumental and exploitative. If some of these ideas about language sound foreign and unfamiliar then let us at least try to write about spirituality, as about all other matters, more carefully, more responsively and with more awareness of the traps that our instrumental, bureaucracy-driven and committee-ridden times set for us.

The attentive reader will, of course, apply the lesson to this chapter.

References

Aristotle (1986) *Nicomachean Ethics*, trans. M. Ostwald, New York: Macmillan.

Armstrong, H. (1976) 'The apprehension of divinity in the self and cosmos in Plotinus', in R. B. Harris (ed.) *The Significance of Neoplatonism*, Norfolk: International Society for Neoplatonic Studies, 187–98.

Armstrong, S. J. and Botzler, R. G. (eds) (1993) *Environmental Ethics: Divergence and Convergence*, New York: McGraw-Hill.

Attfield, R. (1983) *The Ethics of Environmental Concern*, New York: Columbia University Press.

Augustine (1950) *The City of God*, trans. M. Dods, New York: Random House.

Bate, J. (1991) *Romantic Ecology: Wordsworth and the Environmental Tradition*, London: Routledge.

Berry, T. (1991) *Befriending the Earth*, Connecticut: Twenty-Third Publications.

Bhagavadgita (1976) trans. in A. T. de Nicolas, *Avatara: The Humanization of Philosophy through the* Bhagavadgita, Stony Brook: Nicolas Hay.

Bilimoria, P. (1991) 'Indian ethics', in P. Singer (ed.) *Companion to Ethics*, Oxford: Blackwell Reference, 43–57.

—— (1995) 'Duhkha and karma: the problem of evil and God's omnipotence', *Sophia International Journal for Philosophical Theology and Cross-cultural Philosophy of Religion* (100th issue) 34, 1: 92–119.

Bilimoria, P. and Hutchings, P. (1988) 'On disregards for fruits: Kant and the *Gita*', in P. Bilimoria and P. Fenner (eds) *Religions and Comparative Thought*, Delhi: Sri Satguru Publications, 353–67.

Bird-David, N. (1990) 'The giving environment: another perspective on the economic system of Gatherer-Hunters', *Current Anthropology* 31, 2: 189–96.

—— (1992) 'Beyond "The Original Affluent Society": a culturalist reformulation', *Current Anthropology* 33, 1: 25–47.

—— (1993) 'Tribal metaphorization of human-nature relatedness: a comparative analysis', in K. Milton (ed.) *Environmentalism: The View from Anthropology*, London and New York: Routledge.

Blake, W. (1966) *Complete Writings*, ed. G. Keynes, London: Oxford University Press.

Bowers, C. A. (1993) *Critical Essays on Education, Modernity, and the Recovery of the Ecological Imperative*, New York: Teachers College Press.

Briggs, J. and Peat, D. (1989) *Turbulent Mirror*, New York: Harper and Row.

Brightman, R. A. (1987) 'Conservation and resource depletion: the case of the Boreal Forest Algonquians', in B. J. McCay and J. M. Acheson (eds) *The Question of the Commons: The Culture and Ecology of Communal Resources*, Tucson: The University of Arizona Press.

Britton, J. (1970) *Language and Learning*, Harmondsworth: Penguin.

Brownlow, T. (1983) *John Clare and Picturesque Landscape*, Oxford: Clarendon.

Bullock Report (1975) *A Language for Life*, London: HMSO.

Burns, R. (1968) *The Poems and Songs of Robert Burns*, ed. J. Kingsley, Oxford: Clarendon.

Byron, Lord G. G. (1978) *Byron's Poetry*, ed. F. D. McConnell, New York: Norton.

Callicott, J. B. (1982) 'Traditional American Indian and Western European attitudes toward nature: an overview', *Environmental Ethics* 4: 293–318.

—— (1993) 'The land aesthetic', in S. J. Armstrong and R. G. Botzler (eds) *Environmental Ethics: Divergence and Convergence*, New York: McGraw-Hill, 148–57.

Cardenal, E. (1974) *Love*, trans. D. Livingstone, London: Search Press.

Carlson, A. (1993) 'Appreciation and the natural environment', in S. J. Armstrong and R. G. Botzler (eds) *Environmental Ethics: Divergence and Convergence*, New York: McGraw-Hill, 142–7.

Carr, D. (1996) 'Rival conceptions of spiritual education', *Journal of Philosophy of Education* 30, 2: 159–78.

Chang, G. (1977) *Six Yogas of Naropa*, Ithaca: Snow Lion Publications.

Chapple, C. K. (1994) *Non-violence to Animals, Earth and Self in Asian Thought*, Albany: State University of New York Press.

—— (forthcoming) 'Traditionalist and renouncer models: toward an indigenous environmentalism', in L. Nelson (ed.) *Ecological Concern in the Religious Traditions of South Asia*, Albany: State University of New York Press.

China Taoist Association (1995) *Faith and Ecology – Taoism*, Beijing: China Taoist Association and Alliance of Religions and Conservation.

Clare, J. (1986) *John Clare: Selected Poetry and Prose*, eds M. Williams and R. Williams, London: Methuen.

Clark, S. R. L. (1989) *Civil Peace and Sacred Order*, Oxford: Clarendon Press.

—— (1993) *How to Think about the Earth*, London: Mowbrays.

—— (1995) 'Objective values, final causes', *Electronic Journal of Analytical Philosophy* 3: 65–78 (http://www.phil.indiana.edu/ejap/).

—— (1997) 'Platonism and the gods of place', T. Chappell (ed.), *Environmental Metaphysics*, Edinburgh: Edinburgh University Press.

Clifford, J. (ed.) (1986) *Writing Culture*, Berkeley and Los Angeles: University of California Press.

Condren, M. (1995) 'Sacrifice and political legitimation: the production of a gendered social order', *Journal of Women's History* 6, 4 and 7, 1.

Cooper, D. E. and Palmer, J. A. (eds) (1992) *The Environment in Question*, London: Routledge.

Coursey, D. G. (1978) 'Some ideological considerations relating to tropical root crop production', in E. K. Fisk (ed.) *The Adaptation of Traditional Agriculture: Socioeconomic Problems of Urbanization*, Development Studies Centre Monograph 11, Canberra: The Australian National University.

Crick, M. (1982) 'Anthropological field research, meaning creation and meaning construction', in D. Parkin (ed.) *Semantic Anthropology*, London: Academic Press.

Dalai Lama (1996) 'World peace', Lecture, Melbourne, 16 September.

Dandekar, R. N. (1979) *Insights into Hinduism*, Delhi: Ajanta Publications.

de Silva, P. (1990) 'Buddhist environmental ethics', in A. H. Badiner (ed.) *Dharma Gaia: A Harvest of Essays in Buddhism and Ecology*, Berkeley: Parallax Press, 17–27.

—— (1991) 'Buddhist ethics', in P. Singer (ed.) *Companion to Ethics*, Oxford: Blackwell, 58–68.

Devall, W. and Sessions, G. (1985) *Deep Ecology: Living as if Nature Really Mattered*, Salt Lake City: Peregrine Smith.

Diamond, I. and Orenstein, G. F. (eds) (1990) *Reweaving the World: The Emergence of Ecofeminism*, San Francisco: Sierra Club.

Dimitrios, His All-Holiness The Ecumenical Patriarch (1990) *Orthodoxy and the Ecological Crisis*, London: The Ecumenical Patriarchate.

Dinnerstein, D. (1987) *The Rocking of the Cradle and the Ruling of the World*, London: The Women's Press.

Disinger, J. (1983) 'Environmental education's definitional problem', *ERIC/ SMEAC Information Bulletin No 2*, Columbus: ERIC/SMEAC.

Dowman, K. (1994) *The Flight of the Garuda*, Boston: Wisdom Publications.

Dwyer, P. D. (1996) 'The invention of nature', in R. F. Ellen and K. Fukui (eds) *Redefining Nature: Ecology, Culture and Domestication*, Oxford: Berg.

Dyson, F. (1991) *From Eros to Gaia*, New York: Pantheon Books.

Eckersley, R. (1992) *Environmentalism and Modern Political Theory*, London: UCL Press.

Ellen, R. F. (1986) 'What Black Elk left unsaid: on the illusory images of Green Primitivism', *Anthropology Today* 2 (6): 8–12.

—— (1993) 'Rhetoric, practice and incentive in the face of the changing times: a case study in Nuaulu attitudes to conservation and deforestation', in K. Milton (ed.) *Environmentalism: The View from Anthropology*, London and New York: Routledge.

—— (1996) 'The cognitive geometry of nature: a contextual approach', in G. Palsson and P. Descola (eds) *Nature and Society: Anthropological Perspectives*, London and New York: Routledge.

—— (forthcoming) 'Forest knowledge, forest transformation: political contingency, historical ecology and the renegotiation of nature in Central Seram', in T. M. Li (ed.) *Transforming the Indonesian Uplands: Marginality, Power and Production*. A summary is published in Li, T. M. and Uhryniuk, L. (eds) 1995 *Agrarian Transformation in the Indonesian Uplands: Conference Proceedings*, EMDI Environmental Reports 48: 15–17.

Elliot, R. (1994) 'Faking nature', in L. Pojman (ed.) *Environmental Ethics: Readings in Theory and Application*, Boston: Jones and Bartlett, 171–6.

Emery, F. (1981) 'Educational paradigms', *Human Futures* Spring: 1–17.

Flores, A. (ed.) (1960) *Anthology of German Poetry from Hölderlin to Rilke in English Translation*, New York: Doubleday.

Fox, W. (1986) 'Towards a deeper ecology?', *Habitat Australia* 13 (4): 26–8.

Gandhi, M. K. (1959) *My Socialism*, Ahmedabad: Navajivan Publishing House.

—— (1962) *The Teaching of the Gita*, ed. A. T. Hingorani, Bombay: Bharatiya Vidya Bhavan; also as *The Gita According to Gandhi*, trans. Mahadeo Desai, Ahmedabad: Navajivan Publishing House, 1946.

Garrard, G. (1996) 'Radical pastoral?', *Studies in Romanticism* 35, 3 (Fall): 449–65.

Girard, R. (1977) *Violence and the Sacred*, trans. P. Gregory, Baltimore: Johns Hopkins University Press.

Girardet, H. (1992) *The Gaia Atlas of Cities*, London: Gaia Books.

Goldman, A. (1995) *Aesthetic Values*, Boulder: Westview.

Gough, N. (1987) 'Learning with environments: towards an ecological paradigm for education', in I. Robottom (ed.) *Environmental Education: Practice and Possibility*, Geelong: Deakin University Press.

Grundelius, E. (1994) 'Focused analysis – asking the right questions', in *First Steps: Local Agenda 21 in Practice*, London: HMSO.

Gudorf, C. (1992) *Victimization: Examining Christian Complicity*, Philadelphia: Trinity Press International.

Hagness, R. (ed.) *Core Curriculum for Primary, Secondary and Adult Education in Norway*, Oslo: The Royal Ministry of Church, Education and Research.

Hargrove, E. (1989) *Foundations of Environmental Ethics*, Englewood Cliffs, NJ: Prentice-Hall.

Harrison, R. P. (1993) *Forests: The Shadow of Civilisation*, London: University of Chicago Press.

Heidegger, M. (1971) *Poetry, Language and Thought*, trans. A. Hofstadter, London: Harper & Row.

—— (1980) *Being and Time*, trans. J. Macquarrie and E. Robinson, Oxford: Blackwell.

Hegel, G. W. F. (1979) *Introduction to Aesthetics*, trans. T. Knox, Oxford: Clarendon Press.

Hertzberg, Rabbi A. (1986) 'The Jewish declaration on nature', *The Assisi Declarations*, London: WWF.

Hines, J. M., Hungerford, H. R. and Tomera, A. (1986) 'Analysis and synthesis of research on responsible environmental behaviour: a meta-analysis', *Journal of Environmental Education* 18, 2: 1–8.

Holy, L. and Stuchlik, M. (1981) 'The structure of folk models', in L. Holy and M. Stuchlik (eds) *The Structure of Folk Models*, London: Academic Press.

Hume, D. (1965) *Essays: Moral, Literary and Political*, Oxford: Oxford University Press.

Hungerford, H. R. (1983) 'The challenges of K-12 environmental education', paper presented at the First National Congress for Environmental Education Futures: Policies and Practices, University of Vermont, Burlington, VT, in A. B. Sacks (ed.) *Monographs in Environmental Education and Environmental Studies*, vol. 1, SMEAC Information Reference Centre, Cleveland.

Ingold, T. (1994) 'From trust to domination: an alternative history of human–animal relations', in A. Manning and J. Serpell (eds) *Animals and Human Society: Changing Perspectives*, London and New York: Routledge.

—— (1996) 'Hunting and gathering as ways of perceiving the environment', in R. F. Ellen and K. Fukui (eds) *Redefining Nature: Ecology, Culture and Domestication*, Oxford: Berg.

IUCN (1970) *International Working Meeting on Environmental Education in the School Curriculum: Final Report*, September, New York: IUCN.

IUCN, UNEP, WWF (1980) *The World Conservation Strategy*, New York.

IUCN, UNEP, WWF (1991) *Caring for the Earth: A Strategy for Sustainable Living*, Gland, Switzerland.

Jacobs, M. (1995) *Politics of the Real World: Meeting the New Century*, London: Earthscan Publications.

Jacobsen, K. A. (1996) 'Bhagavadgita, Ecosophy T, and Deep Ecology', *Inquiry* 39, 2, June: 219–38.

Jaki, S. (1993) *Is There a Universe?*, Liverpool: Liverpool University Press.

Kant, I. (1952) *The Critique of Judgement*, trans. J Meredith, Oxford: Clarendon Press.

—— (1970) *Kant's Political Writings*, ed. H. Reiss, Cambridge: Cambridge University Press.

Kerr, R. (1792) *The Animal Kingdom or Zoological System of the Celebrated Sir Charles Linnaeus*, London: Murray.

Kroeber, K. (1994) *Ecological Literary Criticism*, New York: Columbia University Press.

Kwok Man Ho, Palmer M. and Ramsay, J. (1994) *Tao Te Ching*, Shaftesbury: Element.

Lasch, C. (1984) *The Minimal Self*, London: Pan.

Lee, K. (1995) 'Beauty for ever?', *Environmental Values* 4: 213–26.

Leibniz, G. (1973) *Discourse on Metaphysics/Correspondence with Arnaud/Monadology*, trans. G. Montgomery, La Salle: Open Court.

Leopold, A. (1949) *A Sand County Almanac: And Sketches Here and There*, New York: Oxford University Press.

Levinas, E. (1969) *Totality and Infinity*, trans. A. Lingis, Pittsburgh: Duquesne University Press.

—— (1985) *Ethics and Infinite: Conversations with Philippe Nemo*, trans. R. A. Cohen, Pittsburgh: Duquesne University Press.

Levine, A. (1961) 'A reminiscence', in *Lahai Roi* (in Hebrew), Jerusalem, 5721.

Levine, M. P. (1994) *Pantheism: A Non-theistic Concept of Deity*, London: Routledge.

Lewontin, R. C. (1993) *The Doctrine of DNA*, London: Penguin.

Lloyd, G. (1980) 'Spinoza's environmental ethics', *Inquiry* 23: 293–311.

Lovelock, J. (1991) *Gaia: The Practical Science of Planetary Medicine*, London: Gaia Books.

—— (1995) *The Ages of Gaia*, second edition, London: Oxford University Press.

Luhmann, N. (1986) *Love as Passion: The Codification of Intimacy*, trans. J. Gaines and D. Jones, Oxford: Polity Press.

—— (1989) *Ecological Communication*, trans. J. Bednarz, Chicago: University of Chicago Press.

Lynch, T. (1996) 'Deep ecology as an aesthetic movement', *Environmental Values* 5: 147–60.

McFarland, T. (1969) *Coleridge and the Pantheist Tradition*, Oxford: Oxford University Press.

Manes, C. (1990) *Green Rage: Radical Environmentalism and the Unmaking of Civilization*, Boston: Little, Brown and Company.

Margulis, L. and Sagan, D. (1995) *What is Life?*, London: Weidenfeld & Nicolson.

Martin, M. (1993) 'Rethinking reverence for life', *Between The Species* 9: 204–13.

Mathews, F. (1991) *The Ecological Self*, London: Routledge.

Midgley, M. (1983) *Animals and Why They Matter*, Harmondsworth: Penguin.

—— (1989) *Wisdom, Information and Wonder: What is Knowledge For?*, London: Routledge.

—— (1992a) 'Philosophical plumbing', in P. Griffiths (ed.) *The Need to Philosophise*, Cambridge: Cambridge University Press.

—— (1992b) *Science as Salvation: A Modern Myth and its Meaning*, London: Routledge.

—— (1995a) 'Duties concerning islands', in R. Elliot (ed.) *Environmental Ethics: Oxford Readings in Philosophy*, Oxford: Oxford University Press, 89–103.

—— (1995b) 'The challenge of science: limited knowledge or a new high priesthood?', in A. Race and R. Williamson (eds) *True to This Earth: Global Challenges and Transforming Faith*, Oxford: Oneworld.

Milton, K. (1996) *Environmentalism and Cultural Theory: Exploring the Role of Anthropology in Environmental Discourse*, London and New York: Routledge.

Muir, J. (1977) *The Mountains of California*, Berkeley: Ten Speed.

Murdoch, I. (1992) *Metaphysics as a Guide to Morals*, London: Chatto & Windus.

Naess, A. (1973) 'The shallow and deep long range ecology movements: a summary', *Inquiry* 16: 10–31.

—— (1990) *Ecology, Community and Lifestyle*, trans. D. Rothenburg, Cambridge: Cambridge University Press.

Naseef, His Excellency Dr Abdullah Omar (1986) 'The Muslim declaration on nature', *The Assisi Declarations*, London: WWF.

NCC (1990) *Curriculum Guidance 7: Environmental Education*, York: National Curriculum Council.

—— (1993) *Spiritual and Moral Development – A Discussion Paper*, York: National Curriculum Council.

Norgaard, R. (1994) *Development Betrayed: The End of Progress and a Coevolutionary Revisioning of the Future*, London: Routledge.

OFSTED (1994) *Spiritual, Moral, Social and Cultural Development*, an OFSTED discussion paper, London: The Office for Standards in Education.

Oxley, J. (1996) 'Endangered Tibet', *Living Now!* 5: 1–23, Melbourne: Whole Person Publications.

Palmer, J. A. (1993) 'Development of concern for the environment and formative experiences of educators', *Journal of Environmental Education* 24, 3: 26–31.

—— (1998) *Environmental Education in the Twenty-First Century*, London: Routledge.

Palmer, J. A. and Suggate J. (1996) 'Influences and experiences affecting the pro-environmental behaviour of educators', *Environmental Education Research* 2, 1: 109–22.

Palmer, M. (1996) *Travels through Sacred China*, London: Thorsons.

Palmer, M. with Breuilly, E. (1996) *Chuang Tzu*, London: Penguin Arkana.

Palmer, M., Ramsay, J. with Kwok Man Ho (1995) *Kuan Yin*, London: Thorsons.

Peat, D. (1996) *Blackfoot Physics*, London: Fourth Estate.

Piper, H. W. (1962) *The Active Universe: Pantheism and the Concept of Imagination in the English Romantic Poets*, London: Athlone Press.

Pirsig, R. M. (1975) *Zen and the Art of Motorcycle Maintenance*, London: The Bodley Head.

Plumptre, C. E. (1878) *General Sketch of the History of Pantheism*, Birmingham: Charles Lowe.

Polanyi, M. (1958) *Personal Knowledge*, London: Routledge.

—— (1958) *The Study of Man*, London: Routledge.

Quarrie, J. (ed) (1992) *Earth Summit '92. The United Nations Conference on Environment and Development*, London: The Regency Press.

Ramsey, J. and Hungerford, H. R. (1989) 'The effects of issue investigation and action training on environmental behaviour in 7th grade students', *Journal of Environmental Education* 20, 4: 29–34.

Rayner, S. (1989) 'Fiddling while the globe warms?', *Anthropology Today* 5, 6: 1–2.

Rees, W. (1996) 'Ecological footprints of the future', *People and the Planet* 5: 2.

Reidy, M. (trans.) (1986) *Religion and Nature Interfaith Ceremony*, London: WWF.

Reynolds, J. (1989) *Self-liberation through Seeing with Naked Awareness*, New York: Station Hill Press.

Rig Veda (Rgveda) (1977), trans. in R. Pannikar, N. Shanta and B. Baumer, *The Vedic Experience Mantramanjari*, London: Darton, Longman and Todd.

Rinpoche, S. (1992) *The Tibetan Book of Living and Dying*, San Francisco: Harper.

Robinson, D. N. (1990) 'Wisdom through the ages', in R. Sternberg (ed.) *Wisdom, its Nature, Origins and Development*, Cambridge: Cambridge University Press.

Robinson, M. (1968) ' "The House of the Mighty Hero" or "The House of Enough Paddy"? Some implications of a Sinhalese myth', in E. R. Leach (ed.) *Dialectic in Practical Religion*, Cambridge: Cambridge University Press.

Robottom, I. (1987) 'Towards inquiry-based professional Development', in I. Robottom (ed.) *Environmental Education: Practice and Possibility*, Geelong: Deakin University Press.

Robottom, I. and Hart, P. (1993) *Research in Environmental Education: Engaging the Debate*, Geelong: Deakin University Press.

Rolston III, Holmes (1986) *Philosophy Gone Wild: Essays in Environmental Ethics*, Buffalo: Prometheus.

Rose, D. B. (1992) *Dingo Makes us Human*, Cambridge: Cambridge University Press.

Routley, R. and Routley, V. (1980) 'Human chauvinism and environmental ethics', in D. Mannison, M. McRobbie and R. Routley (eds) *Environmental Philosophy*, Canberra: ANU Press.

Ruether, R. R. (1993) *Gaia and God: An Ecofeminist Theology of Earth Healing*, London: SCM Press.

Sahtouris, E. (1989) *Gaia: The Human Journey from Chaos to Cosmos*, New York: Pocket Books.

Śaṅkara (1976) *The Bhagavad Gita with the Commentary of Sri Sankaracharya*, trans. A. M. Sastry, Madras: Samata Books.

Saso, M. (1978) *The Teachings of Taoist Master Chuang*, New Haven and London: Yale University Press.

Schama, S. (1995) *Landscape and Memory*, New York: Knopf.

Schmink, M., Redford, K. H. and Padoch, C. (1992) 'Traditional peoples and the biosphere: framing the issues and defining the terms', in K. H. Redford and C. Padoch (eds) *Conservation of Neotropical Forests: Working from Traditional Resource Use*, New York: Columbia University Press.

School Curriculum and Assessment Authority (SCAA) (1996) *Education for Adult Life: The Spiritual and Moral Development of Young People* (Discussion Paper 6), London: School Curriculum and Assessment Authority.

School Curriculum and Assessment Authority (SCAA)/Curriculum and Assessment Authority for Wales (ACAC) (1996) *A Guide to the National Curriculum*, London: School Curriculum and Assessment Authority.

Scottish Office (1993) *Learning for Life: A National Strategy for Environmental Education in Scotland*, Edinburgh: Scottish Office.

Sessions, G. (1977) 'Spinoza and Jeffers on man in nature', *Inquiry* 20: 481–528.

Shiva, V. (1988) *Staying Alive: Women, Ecology and Development*, London: Zed Books.

—— (1994) 'Tripping on misconceptions', *The Times of India*, sunday supplement, 26 April.

Shostakovich, D. (1987) *Testimony: The Memoirs of Dmitri Shostakovich*, London: Faber and Faber.

Snyder, G. (1969) *Earth House Hold*, New York: New Directions.

Songani, K. C. (1984) 'Jaina ethics and the meta-ethical trends', in P. M. Marathe (ed.) *Studies in Jainism*, Poona: Indian Philosophical Quarterly Publications, No. 7, 237–47.

Soulé, M. (1995) 'The social siege of nature', in M. Soulé and G. Lease (eds) *Reinventing Nature: Responses to Postmodernism and Deconstructionism*, Washington: Island Press.

Sources of Indian Tradition, Vol. I (1988), rev. and ed. A. T. Embree, New York: Columbia University Press.

Stiver, D. (1996) *The Philosophy of Religious Language: Sign, Symbol and Story*, Oxford: Blackwell.

Sutherland, W. (1992) *The Rio Treaties of the Global NGO Movement: A Documentary Sourcebook*, London: Adamantine Press.

Suzuki, D. (1973) *Zen and Japanese Culture*, Princeton: Princeton University Press.

Sylvan, R. and Bennett, D. (1994) *The Greening of Ethics*, Cambridge: The White Horse Press.

Takashi Sawano (ed.) (1981) *The Art of Japanese Gardening*, London: Hamlyn.

Tanner, A. (1979) *Bringing Home Animals: Religious Ideology and Mode of Production of the Mistassini Cree Hunters*, St John's, Newfoundland: Institute of Social and Economic Research, Memorial University of Newfoundland.

Taylor, C. (1983) *Philosophical Papers 2: Philosophy and the Human Sciences*, Cambridge: Cambridge University Press.

—— (1989) *Sources of the Self*, Cambridge: Cambridge University Press.

Teilhard de Chardin, P. (1965) *The Phenomenon of Man*, trans. B. Wall, London: Fontana.

Thomas, K. (1984) *Man and the Natural World: Changing Attitudes in England 1600–1800*, Harmondsworth: Penguin.

Thompson, J. (1990) 'A refutation of environmental ethics', *Environmental Ethics* 12: 147–60

Tomlins, B. and Froud, K. (1994) *Environmental Education: Teaching Approaches and Students' Attitudes: A Briefing Paper*, Slough: NFER.

Toynbee, A. and Ikeda, D. (1976) *Choose Life*, London: Oxford University Press.

Turner, V. W. (1967) *The Forest of Symbols: Studies in Ndembu Ritual*, Ithaca: Cornell University Press.

UNCED (1992) *Agenda 21*, United Nations Conference on Environment and Development (The Earth Summit), New York.

UNESCO (1978) *Final Report, Intergovernmental Conference on Environmental Education*, organized by UNESCO in cooperation with UNEP, Tbilisi, USSR.

United Nations (1993a) *Report of the United Nations Conference on Environment and Development, Rio de Janeiro, 3–14 June 1992, Volume I: Resolutions Adopted by the Conference*, New York: United Nations.

—— (1993b) *Report of the United Nations Conference on Environment and Development, Rio de Janeiro, 3–14 June 1992, Volume II: Proceedings of the Conference*, New York: United Nations.

van den Breemer, J. P. (1992) 'Ideas and usage: environment in Aouan society, Ivory Coast', in E. Croll and D. Parkin (eds) *Bush Base: Forest Farm, Culture, Environment and Development*, London and New York: Routledge.

Wackernagel, M. and Rees, W. (1996) *Our Ecological Footprint: Reducing Human Impact on Earth*, Philadelphia: New Society.

Watsuji Tetsuro (1961) *Climate and Culture: A Philosophical Study*, Tokyo: Hokuseido.

Weber, T. (1988) *Hugging the Trees: The Story of the Chipko Movement*, Delhi: Viking Press.

Welker, M. (1994) *God the Spirit: A Force Field of Divine Power and Presence*, Minnesota: Fortress Press.

Wheeler, M. (1979) *The Indus Civilization* (3rd edition), Cambridge: Cambridge University Press.

White, J. (1995) *Education and Personal Well-being in a Secular Universe*, London: University of London Institute of Education.

Wilber, K. (1995) *Sex, Ecology and Spirituality: The Spirit of Evolution*, Boston and London: Shambala.

Williams, R. (1993) *The Country and the City*, London: Hogarth.

Wilson, E. (1992) *The Diversity of Life*, Cambridge, Mass.: Harvard University Press.

Wood, H. W. (1985) 'Modern pantheism as an approach to environmental ethics', *Environmental Ethics* 7: 151–64.

Wordsworth, D. (1991) *Grasmere Journals*, ed. P. Woof, Oxford: Clarendon.

Wordsworth, W. (1974) *Prose Works Vol. III*, eds W. J. B. Oxen and J. W. Smyser, Oxford: Clarendon.

—— (1977) *Home at Grasmere*, ed. B. Darlington, Hassocks: Harvester.

—— (1979) *The Prelude: 1799, 1805, 1850*, eds J. M. H. Wordsworth, M. H. Abrams, S. Gill, London: Norton.

—— (1986) *'The Tuft of Primroses' with other late poems . . .*, ed. J. F. Kishel, London: Cornell University Press.

—— (1987) *Poetical Works*, ed. F. de Selincourt, Oxford: Oxford University Press.

Wordsworth, W. and Coleridge, S. T. (1991) *Lyrical Ballads*, eds R. L. Brett, A. R. Jones, London: Routledge.

World Commission on Environment and Development (1987) *Our Common Future*, Oxford: Oxford University Press.

Wright, D. (1987) *English Romantic Verse*, London: Penguin.

Yeats, W. B. (1962) *Explorations*, London: Macmillan.

Zohar, D. (1990) *The Quantum Self*, London: Flamingo.

Index